THE NERVOUS BODY

THE NERVOUS BODY

An Introduction to the
Autonomic Nervous System
and Behaviour

C. VAN TOLLER
*Department of Psychology,
University of Warwick*

JOHN WILEY & SONS
Chichester · New York · Brisbane · Toronto

iv

Library of Congress Cataloging in Publication Data:

Van Toller, C
  The nervous body.

  1. Neuropsychology. 2. Nervous system, Autonomic.
I. Title. [DNLM: 1. Autonomic nervous system.
2. Behaviour. 3. Psychophysiology. WL600 T651n]
QP360.V38   599'.01'88   78-16758

ISBN 0 471 99703 X (Cloth)
ISBN 0 471 99729 3 (Paper)

Typeset in AM747 Times by Preface Ltd, Salisbury, Wilts
Printed in Great Britain by Unwin Brothers Ltd,
The Gresham Press, Old Woking, Surrey

For
Jane, Maxine, Julian, Daniel and Adam;
and tomorrow

Contents

Preface

The ever increasing range of interests by psychologists has resulted in lecturers having insufficient time to deal adequately with fundamental concepts if they are to reach the more interesting and topical material. This book is designed as an introductory textbook for courses that are concerned with various aspects of the autonomic nervous system. The literature concerning autonomic function is vast and proliferates into most areas of psychology and physiology and I have necessarily been selective in my references. I hope I have managed to give an overview that represents most theories and opinions and that my selection will be considered to have been fair.

Many thanks are due to the many colleagues and students who wittingly or unwittingly contributed towards my understanding of the autonomic nervous system.

Malcolm Rolling, David Harper and Hazel Nicholson helped me draw and label the diagrams, I am grateful for the patient way they began again when the original was changed or modified.

I thank Dr. Everard Thornton for reading most of the manuscript and for his useful comments and suggestions. The book was completed during the year I taught in the Department of Psychology, Bucknell University, Pennsylvania and I would like to thank the Chairman, my friend and colleague, Dr. Roger Tarpy, and the other members of staff for providing a conducive atmosphere for writing.

It is of particular pleasure to formally record my special thanks to my wife, Dr. Mary Van Toller, for drawing figure 3.10 and redrawing figure 2.6 and for her suggestions, help and encouragement at all stages of producing this book.

I am reminded of the saying, "I have picked a bunch of other men's flowers and nothing but the ribbon that binds them is my own". I apologise for the ribbon and errors of commission and omission.

Bucknell University,
May, 1978.

Figures and Tables

xii

CHAPTER 1

Evolutionary and Historical Factors

When we attempt to systematize the broad changing patterns of the evolutionary kaleidoscope we have to consider several interlocking aspects. The environment is changing, sometimes slowly and imperceptibly, but time witnesses the transformation of the primeval swamp into the clinical austerity of the technological age. The living inhabitants of the earth are changing too in their bodily endowment, and in the extent to which these changes either fit them better or indeed handicap them for their existence in the environment in which they find themselves. Both structural and functional adjustments are taking place. The purpose of this book is to consider a particular aspect of the nervous system namely, the autonomic nervous system whose main function is control of the internal vegetative process of the body.

The human body has two main routes for internal control and communication, these are: by nerve cells that extend either directly or via interconnecting nerve cells from the brain to internal or peripheral receptors; or by the release of chemical hormones into the blood distributed by the vascular system. Nerve cells communicate with other nerve cells or receptors by using chemical substances called neurotransmitters. The action of nerve cells is usually local and specific. Hormones are released from various glands in the body and generally, outside the brain, have a widespread action, although they may have local and very specific roles.

During the evolutionary process progressive development of neural tissue occurs which results in the nervous systems in the higher forms of animals being centralized in the cranium as a brain. The autonomic nervous system forms a major part of the peripheral nervous system although, as we shall see, it has extensive central nervous system structures and functions. In addition, it has an extensive hormonal system. The conception of the autonomic nervous system, as distinct and separate from the central nervous system, arises from the earlier anatomical and physiological knowledge that led to certain identifying characteristics being overemphasized. Some authors, in attempts to reduce separation of the nervous system into the brain versus the autonomic nervous system, have used the term 'vegetative nervous system' but this term tends to imply a rather less essential role than turns out to be the case. Certainly, when we consider the behaviour of humans, we must learn to conceive of the nervous system and its various components in an integrated way. However, the autonomic nervous system is an archaic nervous system and some of its peculiarities can be made comprehensible by considering its origins. The autonomic nervous system has two main parts called the 'sympathetic' and the 'parasympathetic'. The sympathetic nervous system consists of nerves issuing from the spinal cord in the chest region. The parasympathetic nervous system consists of two distinct groups of nerves, one set issuing from the cranial region, the other set issuing from the bottom of the spinal cord in the lower back region.

NERVE CELLS AND GANGLIA

The evolutionary tree is conceived as a multibranching process reaching upwards from precellular life to *Homo sapiens* at the apex displaying a complex integrative nervous system. Some parts of our complex nervous system have evolved in comparatively recent times, while other areas are of much older vintage. If we examine the evolutionary process of nervous tissue, we find progressively more freedom of action for the organism arising from the initial development of specialized nerve cells which became organized into increasingly complex nervous networks which allowed for integration in an 'all or nothing' way. The next stage involved the separation of nerve cells into separate entities that communicated by chemical means. This stage introduced a one-way valve system with impulses passing in discrete chunks in a certain direction. To allow for discrete action within local areas, nerve cells became clustered into groups of nerve cell bodies (ganglia) at various points in the organism. Ganglia allow for a concerted and integrative action of impulses from a large number of nerve cells and are not found in the earlier simple one-to-one forms of nervous networks. In terms of evolutionary development ganglionic networks appeared fairly early and this system has proved to be a remarkably successful and durable form of nervous system. The best examples of this system are found in Arthropoda (the largest phylum of the animal kingdom consisting of: spiders, crabs, ants, bees, etc.). Brief consideration of the 'social' behaviour of ants and bees reveals that a ganglionic nervous network allows for sophisticated but rigid patterns of behaviour. It is probable that behavioural stereotypy was one of the primary reasons for the ganglionic nervous system being superseded by nervous systems allowing for greater flexibility in responses.

The point of this seemingly esoteric knowledge about the nerve networks of lower animals is that when we consider the autonomic nervous system we find that large parts of it consist of primitive ganglionic networks. Moreover, the autonomic nervous system appears to have retained some of the properties of the simpler nervous networks. For example, one of the early puzzles that confronted surgeons who attempted to curtail part of the activity of the autonomic nervous system was the paradox that a few days after surgery the small amount of the system remaining was able to produce a response that was equivalent, and sometimes greater, than the original response they were attempting to limit. Murray and Thompson (1957), using electronmicrographic techniques, have demonstrated neural regrowth within five days of the removal of sympathetic nerves. In addition Pick (1970) has presented photographic evidence that small ganglia are present inside individual sympathetic nerves.

Taken together these findings help to explain the rather idiosyncratic nature of autonomic function.

Obviously the fact that a nervous system is phylogenetically old does not necessarily mean that it has retained its original function without modification. However, understanding that the autonomic nervous system possesses certain characteristics of the primitive nervous system helps to explain many contradictions in its structure and function.

CENTRAL AND VISCERAL NERVOUS SYSTEM

The ganglionic nervous system, successful as it was in lower animals, come to be superseded by nervous systems being progressively organized into centralized brains to allow refinement of the elaborate teleoceptive receptors required by animals who monitor the environment. Clearly somatic and visceral factors evolved together and, in all the examples given, somatic and visceral factors interact. The concern of this chapter is the relative contribution of the somewhat neglected autonomic nervous system functions.

It is difficult to decide at what precise point in the evolutionary chain the central and peripheral nervous systems had evolved sufficiently for them to be anatomically distinct. Most authorities consider the tiny marine creature *Amphioxus* to be the link between the vertebrates and lower animals. In this tiny creature a centralized nervous system, enabling the animal to move actively, and a peripheral visceral nervous system can be identified as separate systems. However there is a slightly lower form of life called Urochordata (Tunicata; sea squirts) which has been considered by some authorities to be of particular significance for the development of the autonomic nervous system. Tunicates are a group of marine creatures which in adult form are completely sedentary, feeding by using ciliary currents to draw food towards themselves. However, in their larval stage these primitive creatures possess a well developed central nervous system with a notochord that enables them to swim actively (in primitive vertebrates the primary axial supporting structure and centralized nervous system of the body is called a notochord). This notochord is lost in the adult form of tunicates. Thus, during their immature stage they have a separate centralized and visceral nervous system that regresses when the creature reaches the adult stage resulting in a purely visceral animal.

Romer (1958) has stated that this evolutionary fact is very important for he argues that the adult stage of Tunicata does not, as is normally argued, represent a degenerate form of a more active ancestor but that the ancestors of chordates were sedentary feeders. The free swimming form of larval Tunicata represents an evolutionary breakthrough that for unknown reasons was abandoned in the adult stage. This apparently esoteric point has been claimed by Romer to be of great evolutionary significance for the autonomic nervous system because the basic pattern nervous body for chordates is a dual system having a visceral and a somatic component. In advanced mammals we find an imperfect matching of the somatic and visceral nervous systems.

Elliot (1970), makes a similar point by suggesting that in humans we find great inadequacies in the function of the autonomic nervous system that are to a large degree masked by the superiority of the later developing brain structures. This again suggests that the idiosyncrasies found in both structure and function of the autonomic nervous system occur because of its primitive origin, and further that evolutionary refinement of the autonomic nervous system to a large extent ceased because superiority of the developing brain structures compensated for most of the inadequacies in the autonomic nervous system. These ideas have great significance for the ways in which we view psychosomatic complaints and we will return to consider them in a later chapter.

It must not be thought that *Amphioxus* is a miniature model of the human nervous system. For all its streamlined shape, *Amphioxus* has limited movement and lacks a true brain as well as a parasympathetic nervous system. It is not until the elasmobranchs (cartilaginous fish such as sharks and dogfish) that we find a well developed autonomic nervous system with clearly identifiable sympathetic and parasympathetic nerves. Even then the parasympathetic nervous system of the elasmobranchs is confined to the head region, and these animals lack the lower spinal cord sacral outflow found in the higher mammals (Colin-Nichol, 1952).

So far the primitive nature of the autonomic nervous system has been emphasized, but Botar (1967) has presented the converse case for the autonomic nervous system structure being more sophisticated than some workers have allowed. On the basis of phylogenetic factors Botar argues for a classification giving the autonomic nervous system three main constituent parts which, in order of evolution, are: (1) the enteric nerve-cell system, consisting of nerves in the walls of the digestive system; (2) the visceral part of the nervous system, consisting of secondary systems that participate in the digestive and vegetative functions of the body; and (3) the vascular nerve-cell system, controlling the circulatory system of the body. Botar believes that these main parts differ from one another not only in their phylogenetic development, but also in their ontogenetic, morphological, and physiological properties. If this were confirmed it would indicate that our current views about the structure and function of the autonomic nervous system might be considerably oversimplified in a number of areas of study.

TEMPERATURE REGULATION

In addition to the structural factors considered above there are other evolutionary factors that concern the function of the autonomic nervous system, one of the most important concerns temperature regulation. In general terms we can divide animals into warm blooded and cold blooded creatures. Reptiles and amphibians, which form the bulk of the cold blooded animals, are generally found restricted to the more temperate areas of the world. For example, crocodiles and alligators are successful only in temperate climates; in addition, they require aquatic environments to provide large thermal capacities giving stable temperature conditions. The water they live in prevents these animals overheating during the day or cooling too much at night. Desert lizards, lacking water needed to provide the temperature controlled environment of the larger reptiles, spend a large part of their time behaviourally manipulating their body temperature. As the day progresses lizards slowly warm up and eventually are able to drag themselves into the sun where they lie until they reach a body temperature that allows them to hunt for food. They continue to seek food until this activity causes their body temperature to rise above a critical level and they must then seek shade to allow themselves to cool down. Clearly behavioural manipulation of body temperature is unadaptive for higher forms of life, and mammals have developed endothermic mechanisms to enable them to control their body temperatures by internal means. We find that in humans the autonomic nervous system plays a very important role in temperature regulation of the body.

A critical requirement for internal physiological mechanisms of temperature control is to allow internal body heat to be dissipated against a higher external environmental temperature. Newman (1970) has argued that the conditions found in primitive tropical forests permit an inactive animal to lose metabolic heat by radiation but activity requires two or three times more heat to be dissipated and the problem requires a nervous system that allows this heat to be lost. During the course of evolution, nature has provided a variety of solutions to this problem (for example, animals become nocturnal and are active during the cool of the night), but the ancestors of man developed part of the autonomic nervous system called the sympathetic nervous system for this role. Temperature control in humans is largely brought about by two sympathetic nervous system mechanisms: (1) control of the blood flow through the skin; and (2) sweat glands.

Under conditions where body temperature rises above a critical level the sympathetic nervous system vasodilates or expands the blood vessels in the skin. By this method additional blood reaches the skin and heat is radiated away from the body. When body temperature falls below a critical level the sympathetic nervous system causes vasoconstriction of the blood supply in the skin and this serves to insulate the internal body temperature from a lower external temperature. The warmth that is produced by drinking alcohol on a cold day is due to the alcohol causing vasodilation and is an unadaptive response, for it may result in a slight lowering of the internal body temperature which the sympathetic nervous system is trying to maintain by vasoconstriction of the peripheral limbs and external skin. The other major mechanism of the sympathetic nervous system concerns the sweat glands which become active under very hot conditions. The human body contains from two to five million sweat glands, this is far more than any other primate species. All terrestial mammals have some sweat glands and Weiner and Hellman (1960) have said that humans are not noted for the size or number of their sweat glands but they are for the rate of secretion. Wilson (1975) has stated that it was the sweat glands in conjunction with the loss of body hair that allowed *Homo sapiens* to engage in strenuous pursuit of prey in the African plains.

Most mammals use panting as the main means of temperature control but in man who is so dependent on oral communication this would clearly be unadaptive. It is interesting to speculate whether speech was facilitated in humans by sympathetic mechanisms of temperature control or did the sweat gland system undergo a later period of refinement to allow for speech? One curious fact arises when considering the sweat glands: it is that although these glands are held to be under sympathetic control, the transmitter substance found is not the sympathetic neurotransmitter, noradrenaline, but the main central nervous system neurotransmitter, acetylcholine. Could it be that this part of temperature control was too important and the brain retained direct control over it, or is it an example of a brain-autonomic nervous system interaction that we do not fully appreciate?

'FIGHT OR FLIGHT'

Virtually all of *Homo sapiens* history has been spent in a role consisting predominantly of hunting and food gathering. If archaeological records of humans go back five million years it is within the last 12,000 years that the development of urban and industrial living has occurred. In fact, industrial and technological life is very recent and still not shared to a significant degree by large members of mankind. So, the human body is adapted for movements involved in hunting and other terrestial activities. Its anatomy is adapted for the physical activity and the upright stance is maintained by powerful muscles that also aid locomotion. In the last section loss of body heat was pointed out as probably an important factor in enabling man to hunt.

Cannon (1930) felt that one of the most important roles for the autonomic nervous system was its function in serving to prepare the mammal for emergencies requiring a 'fight or flight' response. When a human encounters a dangerous situation, the autonomic nervous system has the essential role of facilitating bodily actions that help muscular efficiency and reducing bodily functions that do not aid muscular activity. The mechanisms are familiar: non-essential digestive functions cease, resulting in the 'leaden' stomach feeling; saliva flow ceases, producing a dry mouth; the skin blanches as blood is withdrawn from the skin to supplement muscles and, incidentally, reduce bleeding from the surface of the body; sweat glands increase their rate of secretion to dissipate additional body heat and, perhaps, serve to make the skin slippery against blows and attempts to grasp it; heart-rate and respiration quickens; and activation of the sympathetic nervous system causes the pupils of the eyes to dilate which serves to aid the visual processes. Clearly the autonomic nervous system, in particular the sympathetic component, has important roles in response to physical threats and has undergone important adaptive changes to allow for this. One of the consequences of modern industrialized society is that humans must exist in intimate face-to-face contact with limited means of 'burning up' autonomic and hormonal outpourings in physical activity. Social conventions require that civilized humans largely inhibit their muscular responses to emergency and anger stages. Charvat *et al.* (1964) have pointed out that the originally coordinated somatomotor, visceromotor, and hormonal discharge patterns of the body are required to be dissociated by modern conventions, and that the consequences arising from this suppression of muscular activity have yet to be understood in detail. We might also ask what happens to these internal functions when they operate against intellectual and ideological threats rather than the physical ones from which they were originally evolved to protect humans. Modern society tends to result in people having to experience considerable anxiety over extended periods; it has been said that anxiety is fear spread thin. What happens when the bodily defence systems which evolved to operate in emergencies are sustained over the long periods of time required by the stresses and strains of contemporary living. Indeed is this a real problem or has man merely exchanged the very real physical discomforts of his ancestors for the intellectual discomforts of more recent times? Whatever the answer the autonomic nervous system appears to produce many unadaptive responses for urban life (Pick, 1954). This raises the question of the possibility of altering the functions of the autonomic nervous system to make it more compatible with modern society. Do we in fact need to change its structure and function? A few years ago the technique of biofeedback was heralded as the answer to problems concerning the malfunction of the autonomic nervous system; unfortunately, by and large, biofeedback techniques have failed to live up to the earlier expectations. However, consideration of the evolutionary factors in the structure and function of the autonomic nervous system should enlighten us when considering psychosomatic illnesses that derive from diseases of the autonomic system. If nothing more it should make us aware that individual differences in terms of the structure and function of the autonomic nervous system should be considered the rule rather than the exception.

HISTORICAL KNOWLEDGE AND THE UNDERSTANDING OF THE AUTONOMIC NERVOUS SYSTEM

I do not intend to give an exhaustive history of the autonomic nervous system, but rather focus on figures who have played a crucial role, both in the method of enquiry and knowledge about the function of the system. Any reader who wishes to look in more detail at historical developments will find Sheehan (1936) an excellent source.

When we look at present knowledge about the autonomic nervous system we see vestiges of many earlier approaches. Although studies over time have shaded into one another either in a thesis-antithesis fashion, or merely a broadening of complementary earlier knowledge, we can discern five different emphases within the broad endeavour as follows: (1) description; (2) experimental physiology; (3) histology; (4) embryology; and (5) pharmacology. Let us look at each of these stages more closely noting that there is considerable overlap of stages.

Descriptive Stage

Descriptions of the autonomic nervous system actually began with the Greeks, Hippocrates 460-357 B.C. and Aristotle 384-322 B.C., but it is to the later Alexandrian anatomists we are indebted for the first clear descriptions. Galen, A.D. 130-210, was born a Greek, studied in Corinth and Alexandria and later returned to Greece as physician to the gladiators in his home town before finally settling in Rome as physician to the Emperor. The experiments recorded by Galen (Singer, 1956) determined the physiological outlook in medicine that was not to be bettered until Harvey published his important results concerning circulation of the blood in 1628.

Galen, while rejecting the earlier view of Aristotle that the function of the brain was to cool the blood, reinstated an early notion of it as a primary organ generating vital 'animal spirits'. The 'animal spirits' generated in the brain were held to flow along the hollow nerves under the influence of the brain's pulsating action. The idea of a pulsating brain probably arose from analogy with the heart whose pulsating action Galen must have witnessed frequently in his wounded gladiator patients. It is curious to note in passing that although, according to Singer (1956), the Egyptian physicians knew of the relationship between the pulse and blood flow by at least 1500 B.C., the exact relationship between the heart and blood flow awaited Harvey many hundreds of years later. The 'animal spirits' that Galen postulated were held to be responsible for both movement and sensation and there were two main views accounting for this dual role. One held that movement arose from the 'animal spirits' being forced out of the brain, while sensation arose from the backwash from the peripheral parts of the body. The opposing view held that the animal spirits had a fast and slow component, the fast caused muscle movement while the slow mediated sensation. What is important to note here is that both these views contain the idea of integration and reciprocity; ideas that still characterize modern conceptions of the autonomic nervous system.

Galen's conceptions also contained these components but he proposed, in contrast with the two earlier views that have just been mentioned, a concept of *sympathy* for the system, arguing that peripheral nerves were joined together and mutually dependent, the purpose of this union was to promote sympathy between different parts of the body. Galen's concept of sympathy was a view of the body functioning as a harmonious and integrated whole. From dissections Galen was able to identify three swellings (the cervical ganglia, see Figure 2.1) in the neck region. Although he was only able to show connections with the spinal cord (the rami communicantes, grey ramus, and white ramus, see enlarged section in Figure 2.4), he held that the autonomic nervous system was connected to the brain via the cranial nerves. Galen's work gave the autonomic nervous system a status in physiological and behavioural explanations, emphasizing that its primary role was to promote integration, or as he put it 'sympathy' between different parts of the body. One of Galen's demonstrations of sympathy was to show that in a ligated limb both sensation and motor movement were lost. As we will see later the concept of sympathy is still extant today. Galen seems to have

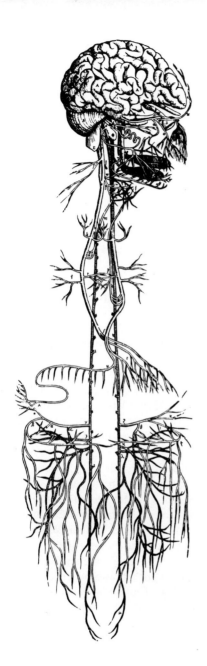

Figure 1.1 Vesalius's drawing of the nervous system of the body. In the drawing reproduced here Vesalius has drawn the whole of the nervous system as it was conceived at that period of time. Vesalius shows the vagus nerve giving off sympathetic nerves at H to the lower left of the drawing. This figure should be compared with Figure 2.1, a modern drawing of the structures of the autonomic nervous system. Reproduced by permission of Harry N. Abrams, Inc.

conceived the autonomic nervous system as a syncytium or interconnecting nervous network in the peripheral part of the body, whose main function was to provide channels for the 'animal spirits'.

For 1,500 years Galen's findings were enshrined as dogma. Medical students were shown dissections not to question and elucidate Galen's thoughts but to enable them to commit this 'orthodox' view to memory. Galen's views survived until the vigour and vitality of the search after 'truth', that characterized the scientific spirit of the Renaissance in the fifteenth century, meant that medical men began to question the earlier dogmas of bodily function. To understand the measure of the power of the Galenical dogma up to this point one has only to look at the drawings of Vesalius (1514-64). These masterly drawings, showing great powers of observation, were inspired by a Latin translation of Galen's works. However, when Vesalius came to draw the autonomic nervous system he produced drawings that must have been at variance with what he observed when he carried out his careful dissections. The differences can be seen by comparing the Vesalian drawing of the autonomic nervous system in Figure 1.1 with Figure 2.1 shown in Chapter 2.

Stage of Experimental Physiology

In the progress of scientific discovery experimental investigation relies upon initial careful descriptions of the phenomena to be investigated and, in general, description and experimentation go hand in hand. In this section, we will consider investigations that marked a turning point in the study of the autonomic nervous system in that they relied heavily on the experimental method.

The impetus for an experimental analysis of the nervous system came most probably from two sources: Galileo's enunciation of the experimental method and its aftermath; Harvey's demonstrations of the circulation of the blood. After Galileo and Harvey the way was opened for systematic and experimental analysis of the nervous system and on account of its accessibility, research on the autonomic nervous system led the way.

Although Descartes (1596-1650) introduced the idea of reflex action occurring without mediation of the central nervous system, during the sixteenth and seventeenth centuries the autonomic nervous system was still conceived as arising directly from the brain. The dual anatomical nature of the autonomic nervous system in the neck region was noticed at this time. Thomas Willis (1621-75), a foremost British physician of his time, accurately described the vagus nerve of the parasympathetic nervous system. The ganglionic swellings of the sympathetic nervous system were conceived of as store houses where the 'animal spirits' were mixed. Willis introduced the idea of involuntary as opposed to voluntary or volitional movement, and argued that the sympathetic nervous system mediated involuntary movement from the cerebrum. He reverted to the Galenical view held that voluntary movement referred to 'animal spirits' from the cerebrum while involuntary movement arose from the cerebellum. Willis produced the most complete and accurate account of the nervous system at that time but he failed to recognize the physiological and anatomical duality of the sympathetic and parasympathetic parts of the autonomic nervous system.

The idea of voluntary and involuntary bodily movements was further extended by Whytt (1714-66), a Scottish physician, who argued that involuntary movements occurred from discrete stimulation of the peripheral nervous system while voluntary movement involved mediation by the brain. Whytt pointed out that movement was normally a mixture of these two aspects. Although not explicitly stating it Whytt came close to stating the concept of the reflex arc (see Figure 1.3). Whytt illustrated the concept of sympathy by pointing to the sympathy that existed between the two eyes of a person having one blind eye and one sound eye. In addition, he suggested that nerves were distinct units arising in the brain and terminating in the peripheral parts of the body. During the period just covered, knowledge concerning the structure and function of the autonomic nervous system was slowly gathered but many of the ideas and concepts were surprisingly modern.

The next group of workers have a common theme in that they discarded the belief held since antiquity of the cerebral origin for the sympathetic nervous system. In 1727, Pourfour du Petit, a French surgeon, made careful dissections of the sympathetic nervous system of dogs and showed that it lacked a cerebral origin. However, it was to be over a century later before this finding was fully accepted and it was concluded that the main connection between the sympathetic and central nervous system was via the spinal connections. Bichat (1771-1802) was responsible for a great impetus to research with his proposal that the autonomic nervous system had a dual somatic and visceral component, and a metabolic role for the system as a whole. Bichat was among the first to visualize the organs of the body as being formed through differentiation of simple functional units or organ precursors. He is also notable in that he differentiated 21 kinds of tissue without the aid of a microscope. To return to our main concern, attention at this time again focused on the ganglia of the sympathetic nervous system. The old idea of pumps mixing 'animal spirits' was replaced by the idea that they acted as miniature brains. The person responsible for this was Johnson who in 1794 drew attention to the fact that ganglia were confined to nerves conducting the involuntary movements (that is the sympathetic nervous system) and the bulk of a ganglion and its outgoing nerves exceeded by many times the volume of nerves and blood supply entering it.

It was during this era that Galvani (1737-98) made the remarkable observation that muscular contractions could be produced in a frog by touching it with two different metals. This technique of electrical stimulation was later to prove a most valuable research tool for discovering the functions of the autonomic nervous system. In 1794 Galvani presented a description of another pioneering experiment with important implications for us here. He caused the muscle on one frog to contract by touching the exposed muscle with the nerve of another frog thus demonstrating for the first time the existence of bioelectrical forces in living tissue. At this point we have reached the beginning of the modern era of scientific research and many important and crucial discoveries were made during the nineteenth century.

Histology

The next stage of development in our understanding of the autonomic nervous system arose from the use of histological techniques that enabled tissues to be hardened and fixed so that they could be sectioned, stained, and examined under microscopes. Initial acceptance of histological techniques was slow, for example, Bichat, whom we have mentioned, was particularly scathing about such techniques and refused to use them or the microscope that had been discovered by Hooke (1635-1703). Bichat maintained that the artefacts produced by histological techniques serve to confuse rather than to clarify understanding. Initially this may have been true, but, with increasing sophistication, a branch of biology grew up that was to be invaluable to our knowledge and understanding of the structure of the autonomic nervous system. Using the newly invented histological techniques and some innovations of their own workers such as Henle, Ranvier, Golgi, Ramon-y-Cajal, and many others were to give their names to cellular and tissue structures that they were the first to observe.

In 1854 Remak published a full and clear account of the structure and connections of the sympathetic ganglia to the spinal cord. He described the lower branch of the rami communicantes as grey and soft in texture while the upper branch was white and of a much firmer texture. At a much later period the observed colour differences were shown to arise from preganglionic nerve fibres being myelinated and surrounded by fatty sheath while the postganglionic nerve fibres lacked the fatty myelin sheath around them (see Figure 2.4). As we shall see later, lack of myelin around the postganglionic fibres results in a slower conduction rate for those neurones. The essential point is that following Remak's studies, workers accepted that the sympathetic nervous system lacked any direct connection to the brain.

The next phase concerned increasing knowledge about the physiological functions of the sympathetic nervous system and its role in controlling the vascular system. Histological techniques had enabled Johannes Müller in 1838 to recognize the differences between the striated muscles of the skeletal system and the non-striated or smooth muscle found in the bladder or the iris of the eye. But, although Henle had shown that sympathetic nerve fibres were distributed along the blood vessels and had suggested they had a motor function, there was still some doubt as to whether the arteries possessed a true muscular coat. It was during this period Claude Bernard (1813-78) began observing the effects of the nervous system on body temperature. In two papers published in 1851 and 1852, Bernard described how blood flow and the temperature of the skin increased on the same side of the head from which he removed the sympathetic nerves. It was Brown-Sequard (1852) who appears to have grasped the full significance of Bernard's observations and suggested that sectioning sympathetic nerves in the cervical region led to paralysis of the vasomotor nerves which produced dilation of blood vessels. Brown-Sequard provided corollary proof by performing an experiment in which he showed that he could constrict blood vessels by electrically stimulating sympathetic nerves. Later in the same year Bernard independently made the same discovery and showed that reduction of blood flow produced a decrease in the temperature of the skin. The essential point was the proof these studies provided of control of the vascular system by the sympathetic nervous system. This control is achieved by controlling the smooth muscles around the blood vessels.

Bernard made two other crucial findings concerning the action of the autonomic nervous system. First, he showed that if he stimulated an area around the fourth ventricle, it resulted in transient sympathetic nervous system effects such as polyuria and glycosuria that were specific to the area being stimulated. Bernard was stimulating areas of the hypothalamus that were later to be shown as important for motivational states. At last, evidence had been found to support the much earlier idea that the autonomic nervous system was at least partially under control of the central nervous system. Bernard's finding stimulated a great deal of interest and research. Second, Bernard introduced the important concept of *milieu interieur*, pointing out that in order to maintain an internal constancy, organisms required protection from the vagaries of the external environment and this was achieved by an organism being encapsulated by its surrounding skin.

Embryology

During this period embryology resulted in a series of discoveries concerning the origin of the sympathetic nervous system. It was shown that primordia of the sympathetic nervous system arose from the same neural crest cells, which form the brain. Certain neural crest cells migrate to positions alongside the vertebral column where they form the two parallel chains of the paravertebral ganglia (see Figure 1.2). Other neural crest cells migrate into the visceral cavity to form the prevertebral ganglia in the visceral and abdominal regions. Finally, some neural crest cells migrate to take up position at the apex of the adrenal gland where, encapsulated by cells that form the adrenal cortex, they form the adrenal medulla whose primary function is to produce chemicals vital to the function of the autonomic nervous system. We will consider the adrenal medullary and adrenal cortex cells in greater detail in a later chapter, but note at this point that the cells of the adrenal cortex and medulla are embryologically distinct because the cells of the adrenal cortex do not arise from neural crest cells.

By the end of the eighteenth century knowledge about the function of the autonomic nervous system indicated that it was very important for maintaining normal visceral functions and that it had an important role in regulating blood flow. The idea that it was completely independent of the central nervous system had been shown to be incorrect. The white and grey rami communicantes were recognized as the link with the central nervous system but it was believed that the sympathetic nervous system possessed nerve fibres that passed back into the spinal cord providing feedback to the brain. Up to the present time such nerve fibres have not been unequivocally demonstrated. The exact nature of the reciprocal interaction of the sympathetic and parasympathetic nervous systems was not fully recongnized despite a finding made during the century that the vagus nerve had an inhibitory effect on heart-rate.

If we turn our attention now to work done from the middle of the eighteenth

14

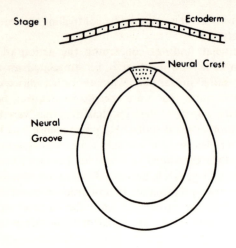

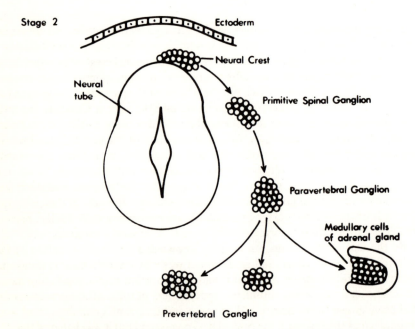

Figure 1.2 Embryological development of the sympathetic ganglia. Stage 1 shows the neural crest forming on the dorsal surface of the neural groove. Stage 2 shows cells from the neural crest migrating to form the primitive spinal ganglia of the sympathetic nervous system that come to lie alongside the spinal cord. Some of the cells from the neural crest continue into the abdominal cavity where they come to form the prevertebral ganglia. Other cells migrate to the apex of the kidney where they become encapsulated as the adrenal medullary cells by cells that are forming the adrenal cortex. Reproduced by permission of the Longman Group Ltd.

century, the next major worker of note was Gaskell (1847-1914) whose researches extend into this century. Gaskell's initial studies consisted of careful observations of stained serial sections through the ventral and dorsal spinal roots. He concluded that the only possible link between the sympathetic nervous system and the central nervous system was via the rami communicantes and his observations led him to state the concept of a reflex neuronal arc consisting of three basic nerve cells (see Figure 1.3). The somatic or striated muscle neurone arc consists of a sensory or receptor nerve cell from the peripheral tissues entering the spinal cord via the dorsal root. A second nerve cell, the internuncial nerve cell, provides a connection between the sensory nerve cell which starts in the ventral horn of the spinal cord. The third motor nerve cell leaves the spinal cord via the ventral root and innervates an appropriate skeletal muscle unit. The reflex arc of the sympathetic nervous system, which differed in the neurone of the internuncial nerve cell, travels out of the spinal cord via the ventral root and terminates in one or other of the twin chains of paravertebral ganglia alongside the vertebral column.

It should be realized that it is very doubtful if reflex arcs are ever encountered in the simple one-to-one schematic form described above. However, it was the concept of the reflex arc, perhaps more than any other single idea, which made the structure and function of the autonomic nervous system comprehensible.

In 1885, Gaskell proposed that tissues might be innervated by two mutually antagonistic nervous systems; however, the proof of his suggestion was provided by Langley whose work overlapped with Gaskell's research. Langley (1852-1925) provided the most complete modern statement about the structure and function of the autonomic nervous system. He was responsible for the term autonomic nervous system and he understood the reciprocal relationship between cranial and sacral outflows of the parasympathetic nervous system and the thoraco-lumbar outflows of the sympathetic nervous system. Detailed consideration of these points will be taken up in a later chapter but the basic points that are to be made here are threefold. First, the sympathetic nervous system's action is diffuse and it is confined to the neck and chest regions. Second, the parasympathetic system is more discrete and arises as two separate outflows, one from the head and the other from lower back regions of the vertebral column. Finally, in general terms, both systems are found to innervate organs and act in a reciprocal way. How Langley plotted the structure of the autonomic nervous system will be explained in the next section dealing with early pharmacological knowledge. Langley proposed the term 'autonomic' and was aware the term implied rather more freedom than the system actually possessed. This is shown in a rather paradoxical passage taken from an address to the Royal Society in 1898:

'the word "autonomic" does suggest a much greater degree of independence of the central nervous system than in fact exists, except perhaps in that part which is the walls of the alimentary canal. But it is, I think, more important that new words should be used for new ideas than that the word should be adequately descriptive.'

(A) Somatic nervous system reflex arc

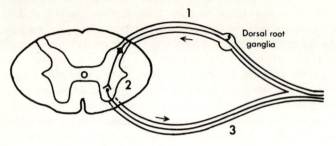

(B) Autonomic nervous system reflex arc

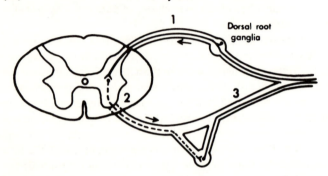

Figure 1.3   The reflex arcs of the somatic and autonomic nervous systems. The reflex arc of the somatic nervous system is conceived as having three neurones. The receptor or sensory nerve cell (1) passes from the surface of the body into the dorsal horn of the spinal cord. Within the spinal cord the first cell synapses with a internuncial cell (2) that passes down to the ventral horn. The final effector cell (3) passes out of the spinal cord to innervate a muscle group in the peripheral part of the body. The reflex arc of the sympathetic nervous system is different. The initial nerve cell (1) in the arc enters the spinal cord at the dorsal horn and synapses with the second cell (2) that passes out of the spinal cord and enters one of the paravertebral ganglia that lie alongside the spinal cord. The second nerve cell synapses within the paravertebral ganglia with a large number of nerve cells (3) that travel out to autonomic receptors. Some of the nerve cells forming the second link (2) in the arc travel through the paravertebral ganglia without synapsing and continue on to innervate the prevertebral ganglia or the adrenal medullary cells.

In the light of Langley's actual passage it is unfortunate that too many subsequent workers have, apparently, used the term autonomic in its most literal sense.

Langley's work can be summarized by saying he suggested the autonomic nervous system was an efferent outflowing nervous system that could be classified as a dual system from three major points: (1) it was anatomically distinct; (2) the sympathetic and parasympathetic neurones' components had opposing physiological actions; and (3) its physiological actions could be mimicked by the action of two distinct classes of drugs. This final point has not been dealt with yet as it concerns early discoveries made in the developing field of psychopharmacology.

Pharmacology

Towards the end of the nineteenth century physiological, anatomical knowledge and theoretical concepts of the autonomic nervous system, in the areas of anatomy and physiology, had advanced as far as they possibly could. Further understanding required additional discoveries from the new and rapidly developing field of pharmacology. This pharmacological information provided the additional conceptual knowledge needed to achieve a comprehensive understanding of the structure and function of the autonomic nervous system.

In 1877 Dubois Reymard had suggested that chemical transmission occurred in the nervous system, but his idea was not generally accepted at that time. However, the idea slowly germinated, and it was only a matter of time before sufficient evidence had been accumulated for the idea to receive serious consideration. Histological studies made at the close of the eighteenth century, by the Spanish anatomist Ramon-y-Cajal, had shown the neurone to be a single entity separated from other nerve cells, and he provided the structural basis for the nervous system. Connections between two adjacent neurones were postulated to occur through their close proximity. It was Sherrington who proposed the name synapse for the gap that was shown to exist between adjacent nerve cells; he also proposed chemical transmission between nerve cells. In doing so Sherrington provided a functional explanation for Ramon-y-Cajal's proposed nervous system structure. Turning upwards around many spirals in the ascent of human knowledge we arrive at a position that could be held to be exactly above Galen's original concept of a nervous syncytium of hollow tubes conveying vital spirits, to a modern concept of an interlinked nervous system with complex neurochemical transmission at the synapses.

Hirschmann (1873) showed that after a moderate dose of the alkaloid drug nicotine, electrical stimulation of a cervical ganglion failed to produce normal pupillary dilation of the eye. Later it was shown that if ganglia were painted with a solution of nicotine, and then stimulated with an electrical current, an initial excitatory phase gave way to a complete inhibition of the ganglia. It was this pharmacological property of nicotine that enabled Langley to systematically coat individual sympathetic ganglia with nicotine and observe the effects of preganglionic stimulation. By this pharmacological technique he was able to map the distribution of sympathetic fibres. For example, by blocking the cervical ganglia with nicotine, Langley was able to show that the dilator muscles of the iris in the eye had sympathetic nervous system connections.

A further breakthrough came in 1897 when Abel synthesized the hormone epinephrine that was released from the adrenal gland. Later, Takamine extracted the adrenal hormone in a pure form and named it adrenaline. (The terms adrenaline and noradrenaline are synonymous with epinephrine and norepinephrine respectively. The former, having a Latin derivation, appear to be favoured by Europeans; the latter, having a Greek derivation, are favoured by the Americans.)

Elliot (1905), while still a student at Cambridge, noted that the effects of administering adrenaline and the effects of electrical stimulation upon the sympathetic nervous system were similar. He suggested that if adrenaline were considered to be a chemical transmitter substance it would explain the similarity of action. This idea was not generally accepted but Dale (1914), noting the similarity between the administration of the drug acetylcholine and electrical stimulation of the parasympathetic nervous system, drew a similar conclusion that acetylcholine was a chemical transmitter substance. He provided additional experimental evidence showing that the action of acetylcholine on smooth muscle was blocked by the drug atropine and potentiated by the drug eserine. He further suggested that acetylcholine was a chemical that mimicked or imitated parasympathetic effects. The results of Elliot's and Dale's observations indicated that the sympathetic and parasympathetic nervous systems had separate chemicals mediating their synaptic transmissions. Today we would speak of acetylcholine as the neurotransmitter substance for the parasympathetic nervous system and adrenaline as the neurotransmitter substance for the sympathetic nervous system. As we shall see later, adrenaline is not the transmitted substance for the sympathetic nervous system but this fact was not known until the late 1940s; we still use the term adrenergic neurone for sympathetic nerve fibres. Much later Dale (1933) proposed that it would be of value to have a pharmacological concept to explain certain actions of the autonomic nervous system. He proposed that nerve fibres releasing an adrenaline-like substance be called *adrenergic* and nerve fibres releasing an acetylcholine-like substance be called *cholinergic*.

The crucial piece of pharmacological evidence came from Loewi (1921) who noted that the duration of inhibition in a beating heart, produced by electrical stimulation of the vagal nerve, was considerably longer than the duration of the electrical stimulation. He proposed that the explanation for this phenomenon might lie in a humoral substance. His experimental proof of the existence of a humoral agent involved collecting a perfusate from an isolated frog's heart that was being electrically stimulated via the vagal nerve, and demonstrating that the perfusate by itself was able to slow the heart-rate of a second isolated frog's heart. Loewi also showed that the perfusate collected from an isolated frog's heart undergoing stimulation of the sympathetic nerves was able to speed up the heart-rate of a second frog's heart. With this finding there could be very little doubt that neurochemicals played a crucial role in the function of the autonomic nervous system.

The final stages of our modern comprehension of the autonomic nervous system's neurochemistry was nearly complete, but there was one final problem that proved rather difficult to solve: it concerns Dale's earlier suggestion that adrenaline was the neurotransmitter substance for the sympathetic nervous system. Throughout the 1920s and 1930s it was assumed that the substance released from the adrenal medullary cells producing sympathetic activity was adrenaline. However, evidence slowly accumulated to show that the active substance, although chemically very similar to adrenaline, was not adrenaline. For example, at the same time as Loewi was carrying out his experiments on the frog's heart, Cannon and his collaborators were carrying out a series of studies designed to elucidate the role of the sympathoadrenal system. Cannon's main technique was to denervate the heart of an animal to observe the hormonal effects produced by stimulating the splanchnics leading to the adrenal gland (see Figure 2.6). From his studies Cannon argued that the sympatho-adrenal system was an integrated system. He discovered that the adrenal substance that stimulated the sympathetic system was also released from the nerve endings of the sympathetic nerves. He called the adrenal substance *sympathin* recognizing that it was different from adrenaline but he did not anticipate the later findings concerning the significance of noradrenaline. This discovery was delayed because noradrenaline is a precursor, or earlier stage compound, produced in the metabolism of adrenaline.

The problem was finally resolved by von Euler (1946) who showed that extracts from sympathetic nerves innervating the adrenal gland had similar characteristics to the hormone noradrenaline. He suggested that the neurotransmitter substance for the sympathetic nervous system was noradrenaline. Confirmation of von Euler's conclusion was provided by Peart (1949) who demonstrated that stimulation of the splenic nerves caused noradrenaline to be released. The irony was that noradrenaline had been known since the earliest part of the twentieth century, but the fact that it was a precursor of adrenaline (see Figure 3.2) delayed recognition that it had a completely independent role. The independence of noradrenaline as a transmitter substance will be considered in Chapter 2 where the roles of adrenaline and noradrenaline (the catecholamines) are considered in greater detail.

No historical discussion of the autonomic nervous system would be complete without reference to the researches of W.B. Cannon and his collaborators, his concepts are central to any discussion of the structure and function of the autonomic nervous system and behaviour. Cannon's researches and conclusions are summarized in four monographs written over a period of 20 years (Cannon, 1929; Cannon 1932; Cannon and Rosenblueth, 1937; Cannon and Rosenblueth, 1949).

Following Claude Bernard's earlier concept of the *milieu interieur*, Richet (1900) indicated that a further refinement of this concept was required when, referring to the living body, he wrote:

'By apparent contradiction it maintains its stability only if it is excitable and capable of modifying itself according to external stimuli and adjusting in response to the stimulation.'

Cannon demonstrated that this particular internal function was one of the essential roles of the autonomic nervous system. He used the term homeostasis to mean the maintenance of essential bodily functions such as temperature, acidity/alkalinity, blood pressure, etc., altering within certain limits around critical mean values. This concept has been of great value in the fields of physiology and biology.

Cannon did much to clarify ambiguities and contradictions concerning the structure and function of the autonomic nervous system. He was responsible for formulating the concept of the sympathetic nervous system as a catabolic or spending system and the parasympathetic nervous system as an anabolic or conserving system. He showed that it was possible to remove totally the sympathetic nervous system (sympathectomy) from laboratory animals. Sympathectomized animals were invaluable in elucidating the physiological roles of the autonomic nervous system. In addition, he carried out a series of studies to discover the effects on behaviour following sympathectomy. Cannon's concept of the role of the sympathetic nervous system centred around its function in emergency states and this gave rise to the concept of 'fight or flight'.

Finally, Cannon and Rosenblueth (1937) examined the problem which related to the ability of the autonomic nervous system to compensate for curtailment or reduction of its activity. As indicated earlier in this chapter, observations following surgery had drawn attention to this phenomenon pointing out that in many patients there were actually an increase in autonomic activity. Cannon and Rosenblueth called the phenomenon supersensitivity and illustrated their account with the physiological paradoxes it could produce. The ability of the autonomic nervous system to compensate appears to be widespread, Trendelenburg (1966) has reviewed the physiological evidence, and Emmelin (1961) the pharmacological evidence. Apart from these findings the reader may recall the finding of Murray and Thompson (1957) of neural regrowth within very short postoperative periods.

REFERENCES

Botar, J. (1967) 'Phylogenetic evolution of the vegetative nervous system', *Acta Neurovegativa (Wien)* **30**, 342-54.

Cannon, W.B. (1929) *Bodily Changes in Pain, Hunger, Fear and Rage*. Appleton, New York.

Cannon, W.B. (1930), 'The autonomic nervous system: an interpretation. The Linacre lecture, 1930', *Lancet*, 1109-15.

Cannon, W.B. (1932) *The Wisdom of the Body*. Appleton-Century-Crofts, New York.

Cannon, W.B. and Rosenblueth, A. (1933) 'Sympathin E and sympathin I', *Amer. J. Physiol.*, **104**, 557-74.

Cannon, W.B. and Rosenblueth, A. (1937) *Autonomic Neuro-effector Systems*. Macmillan and Co., New York.

Cannon W.B., and Rosenblueth, A. (1949). *The Supersensitivity of Denervated Structures*. Macmillan, New York.

Charvat, J., Dell, P. and Folkow, B. (1964) 'Mental factors and cardiovascular disorder', *Cardiologi*, **44**, 124-41.

Colin-Nicol, J.A. (1952) 'Autonomic nervous systems in lower chordates', *Biological Review. Cambridge Philosophical Society* **27**, 1-49.

Dale, H.H. (1914) 'The action of certain esters and ethers of choline, and their relation to muscarine', *J. Pharmacol.*, **6**, 147-90.

Dale, H.H. (1933) 'Nomenclature of the fibres in the autonomic nervous system and their effects', *J. Physiol.* (London), **80**, 10p.

Elliott, H.C. (1970) *The Shape of Intelligence: The Evolution of the Human Brain.* Allen and Unwin, London.

Elliott, T.R. (1905) The action of adrenaline. *J. Physiol.* (London), **32**, 401-67.

Emmelin, N. (1961) 'Supersensitivity following "pharmacological denervation"', *Pharmacol. Rev.*, **13**, 17-38.

Euler von, U.S. (1946) 'A specific sympathomimetic ergone in adrenergic nerve fibres (sympathin) and its relations to adrenaline and noradrenaline', *Acta Physiologia Scandinavia*, **12**, 46-73.

Loewi, O. (1921) 'Uber humorale ubertragbarkeit der herznervenwi rkung', *Pflugers Achives ges Physiologie*, **189**, 239-42.

Murray, J.G. and Thompson, J.W. (1957) 'Collateral sprouting in response to injury of the autonomic nervous system, and its consequences', *Brit. Med. Bull.*, **13**, 213-9.

Newman, R.W. (1970), 'Why man is such a sweaty and thirsty naked animal: a speculative view', *Human Biology*, **42**, 12-27.

Peart, W.S. (1949) 'The nature of splenic sympathin', *J. Physiol.* **108**, 491-501.

Pick, J. (1954) 'The evolution of homeostasis: the phylogenetic development of the regulation of bodily and mental activities by the autonomic nervous system', *Proc. Amer. Philosophy Society*, **98**, 298-303.

Pick, J. (1970) *The Autonomic Nervous System: Morphological, Comparative, Clinical and Surgical Aspects.* Lippincott, New York.

Richet, Ch. (1900) *Dictionnaire de Physiology.* Alcan, Paris.

Romer, A.S. (1958) 'Phylogeny and behaviour with special reference to vertebrate evolution', in A. Roe and G.C. Simpson (Eds.) *Behaviour and Evolution.* Yale University Press.

Sheehan, D. (1936) 'Discovery of the autonomic nervous system', *Arch. Neurol. Psychiat.*, **35**, 1081-115.

Singer, C. (1956) *Galen on Anatomical Procedures.* Oxford University Press, London.

Singer, C. (1956) *The Discovery of the Circulation of the Blood.* Dawson, London.

Trendelenburg, U. (1966) 'Denervation supersensitivity of structures innervated by the autonomic nervous system', *Acta Cientifica Venezolana*, **17**, 138-42.

Weiner, J.S. and Hellmann, (1960) 'The sweat glands', *Biol. Rev.*, **25**, 141-86.

Wilson, E.O. (1975) *Sociobiology: The New Synthesis.* Belknap Harvard Uni. Press, Mass.

CHAPTER 2

The Peripheral Autonomic Nervous System

Having briefly sketched important evolutionary and historical aspects we can now consider the autonomic nervous system as it is understood at the present time. The peripheral autonomic nervous system consists of a network of nerves which, together with associated hormones, have a motor control over the various forms of smooth muscle found in the abdominal region and around the blood vessels. It also controls the piloerector muscles to the hairs on the body and the numerous sweat glands that are found in certain parts of the body. Many textbooks give the impression that the central nervous system and the autonomic nervous system are separate entities, indeed the schematic drawings shown in Figures 2.1, 2.2 and 2.3 might be held to contribute to this belief but it is important to understand that the separations shown are given to aid conceptualization. It must be remembered that separation of the nervous system into two parts arose from earlier conceptions of what was understood about its anatomy and function rather than reflecting two fundamental parts of the nervous system found inside the body. Indeed, previously increased comprehension came from conceiving the system as quite separate but it is now more important to conceive the nervous system as an integrated whole and to tease out the subtle interactions that occur between the brain and the autonomic nervous system. In a small attempt in this direction Figure 2.1 indicates brain areas that are known to have important autonomic roles, but detailed discussion of these points will not be made until Chapter 3.

The peripheral parts of the autonomic nervous system are divided into the parasympathetic nervous system and the sympathetic nervous system. The parasympathetic nervous system leaves the central nervous system at two locations: the cranial parasympathetic nerves leave the central nervous system at the base of the brain with other cranial nerves, and the sacral parasympathetic nerves leave the central nervous system at the bottom of the spinal cord. Sympathetic nerves leave the central nervous system between the cervical and thoracic regions of the spinal cord; this is, the neck and chest region of the body.

Sensorimotor nerves leave the central nervous system via the spinal cord and innervate muscles without interruption. In contrast, the nerves of the autonomic nervous system, in particular the nerves of the sympathetic system, are influenced by many other autonomic nerves. The difference between the central nervous system motor nerves and the autonomic nervous system motor nerves is illustrated in Figure 2.2. Also note that the cell bodies of the autonomic nerves

Figure 2.1 A schematic drawing of the autonomic nervous system. This diagram shows the brain and peripheral components of the autonomic nervous system. The letters and numerals to the left of the left-hand section of the spine refer to the cervical, thoracic, lumbar, and sacral vertebrae and the sympathetic nerves issuing from them. The important brain areas and neuroendocrine glands of the brain are indicated on the medial plane of the right hemisphere of the brain. On the left-hand side of the diagram the nerve fibres of the sympathetic nervous system are shown issuing from the thoracolumbar regions of the spinal cord. The nerve fibres of the parasympathetic nervous system are shown issuing from the cranial and sacral ends of the right-hand spinal cord. A key to the pre- and postganglionic nerve fibres is given in the box below the diagram. The names of the prevertebral ganglia are shown in the key at the bottom right-hand side of the diagram. Adapted from Guyton, *Structure and Function of the Nervous System*, W. B. Saunders Company, 1972.

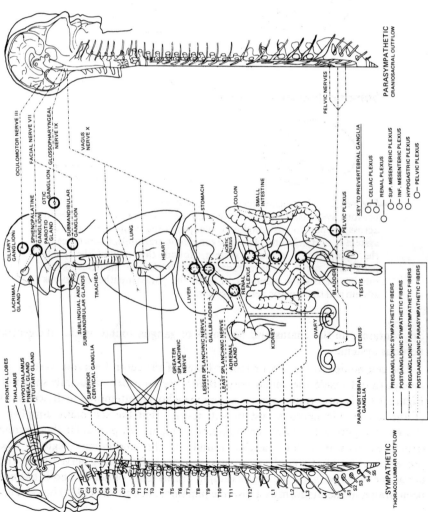

Sympathetic Nervous System

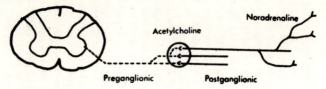

Noradrenaline

Acetylcholine

Preganglionic   Postganglionic

Parasympathetic Nervous System

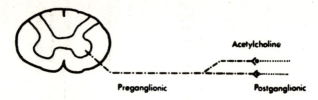

Acetylcholine

Preganglionic   Postganglionic

Somatic Nervous System

Acetylcholine

Figure 2.2   The transmitter substances of the somatic and autonomic nervous systems. The transmitter substance mediating transmission across the pre- to postganglionic synapse of the sympathetic nervous system is acetylcholine. The postganglionic transmitter substance of the sympathetic nervous system, found throughout the post-ganglionic nerve, is noradrenaline. Acetylcholine mediates transmission across the synapses of both the parasympathetic nervous system and the somatic nervous system.

are contained in the chains and networks of ganglia found outside the central nervous system.

The cell bodies of the sensorimotor fibres of the central nervous system are always found inside the spinal cord. Before we consider the details shown in Figure 2.1 it is important to note, as anyone who has attempted to dissect parts of the autonomic nervous system will testify, that the schematic simplicity shown in Figure 2.1 is never encountered in reality. In particular the discrete structures indicated for ganglia found in the abdominal regions (the prevertebral ganglia) take the form of a mass of anastomosing nerve fibres. The aortic plexus is perhaps the best example of this mass of nerves and it has not been drawn in the diagram as a discrete junction. Jacobowitz (1974) has given a

25

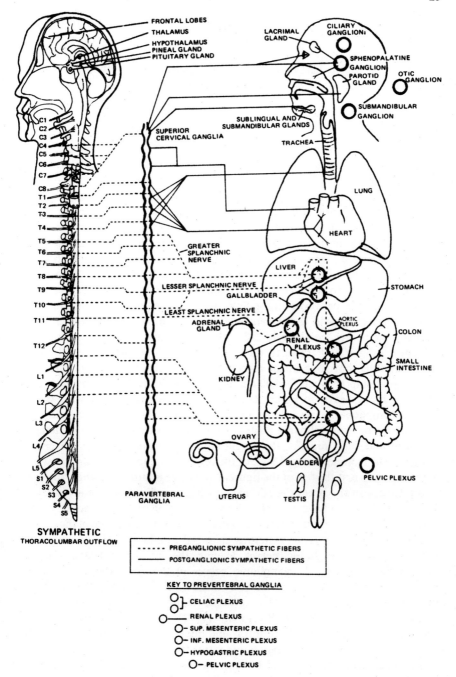

Figure 2.3   A schematic drawing of the sympathetic nervous system. This figure shows the sympathetic innervation described in more detail in Figure 2.1. Adapted from Guyton, *Structure and Function of the Nervous System*, W.B. Saunders Company, 1972.

good recent account which includes a number of micrographs showing the structure of ganglia.

Despite being schematic, the drawings shown in Figure 2.1 are still very complex and will be best understood if you break your search into small units and in addition work backwards between Figures 2.1, 2.2 and 2.3.

THE SYMPATHETIC NERVOUS SYSTEM

The sympathetic system is shown in Figure 2.3 and, using the key given at the foot of the diagram showing the types of nerve fibre, you can trace the routes of the pre- and postganglionic nerves. The letters and numerals shown alongside the vertebral column refer to the separate vertebral sections of the spinal cord. In descending order these are: cervical, C1-C8; thoracic, T1-T12; lumbar, L1-L5; sacral, S1-S5. The nervous outflows at these points are labelled accordingly. The myelinated fibres of the preganglionic nerves leave the spinal cord and enter one of the paravertebral ganglia and do one of three things: (1) connect with a nerve cell of an unmyelinated post-ganglionic fibre; (2) ascend or descend the ganglionic chains to connect with postganglionic nerves at another level; or (3) pass through the paravertebral ganglia without synapsing and run out to the prevertebral ganglia in the visceral and gut region. A group of preganglionic sympathetic nerves, collectively called the splanchnics, pass through the paravertebral ganglia and travel to the adrenals where they innervate the adrenal medullae cells that contain the catecholamines adrenaline and noradrenaline. The relationship between the splanchnic nerves and the adrenal medullary cells is shown more clearly in Figure 2.6. The clearest relationship between the pre- and postganglionic nerves is best shown in the cervical and thoracic region from C4-T4 with the unmyelinated postganglionic nerves travelling out to innervate the head and chest regions. Preganglionic nerves running to the prevertebral ganglia issue from segments T5-L2 of the vertebral column. The important brain areas are shown at the top of the figure. Note three important glands, the pineal and pituitary in the brain the adrenal which is found on top of the kidney. All have been shown to have critical roles in the function of the autonomic nervous system. The key to the prevertebral ganglia is shown at the bottom of the figure.

The sympathetic nervous system is arranged to allow maximum interaction between the preganglionic nerves leaving the spinal cord and the final postganglionic nerves. Preganglionic nerves of the sympathetic nerves have many collateral pathways and show a considerable increase in the numbers of pre- to postganglionic nerves. The ratio for the pre- to postganglionic nerves is usually given as I:32 but Blackman (1974) has reported that there is considerable variation for both intra- and interspecies comparisons. Ebbesson (1968), obtained pre- to postganglionic ratios for the superior cervical ganglion of I:28 in squirrel monkeys and I:196 in humans. In addition, Boyd (1957) has reported the existence of small intermediate sympathetic ganglia.

As shown in Figure 2.4 the preganglionic nerve fibres enter the paravertebral ganglia via the white rami communicantes and leave the ganglia via the grey

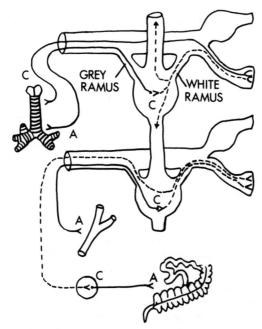

A (NOR) ADRENERGIC ; NORADRENALINE
C CHOLINERGIC ; ACETYLCHOLINE

Figure 2.4 An enlarged drawing showing the sympathetic ganglia and nerves. Preganglionic nerves leave the spinal cord and enter a ganglion as white myelinated nerves in the white rami communicantes and leave via the grey rami communicantes as unmyelinated nerves. The preganglionic neurotransmitter substance is acetylcholine and the postganglionic transmitter substance is noradrenaline. Note that when a sympathetic nerve travels through a ganglion without synapsing it remains a cholinergic nerve. An example of such a sympathetic nerve would be the splanchnics or a sympathetic nerve running out to a sweat gland. Adapted from Guyton, *Structure and Function of the Nervous System*, W.B. Saunders Company, 1972.

rami communicantes. The colour difference between the incoming and outgoing nerves is due to the posganglianic nerve being unmyelinated and not surrounded by a fatty sheath. The absence of the myelin sheath surrounding the nerve results in a slower transmission rate for the postganglionic nerves. It has long been the general view that the ganglia play no significant role in modifying the activity of the autonomic nervous system and that integration is carried out inside the central nervous system. Blackman (1974) has argued that this view, deriving from Langley, should be modified as autonomic ganglia have been shown to be capable of different kinds of activity.

THE PARASYMPATHETIC NERVOUS SYSTEM

The parasympathetic nervous system is found as two discrete neural outflows at the opposite ends of the spinal cord, see Figure 2.5. The cranial nerves of the parasympathetic system largely relate to protective reflexes such as contracting the pupil of the eye against excessive light. The parasympathetic cranial nerves arise from nuclei in the brain stem and travel as preganglionic neurones to small ganglia that are found in or very near to the innervated organ. The III (oculomotor) cranial nerve has the most specific functions controlling protective reflexes of the eye. The VII (facial), IX (glossopharyngeal), and X (vagus) cranial nerves arise from the bulbar region of the brain stem. These nerves control functions in the head and face area but also have branches that innervate viscera in the thorax and abdomen. The VII (facial) cranial nerve has two main pathways that serve the lacrimal and salivary glands but this nerve also has pathways that run to structures of the inner ear. The IX (glossopharyngeal) cranial nerve runs out to the mouth and throat region but has important branches to the heart and corotid sinus (the corotid sinus contains pressure-sensitive receptors which, when stimulated, cause slowing of the heart-rate, vasodilation, and a fall in blood pressure). The X (vagus) cranial nerve is the largest of the parasympathetic nerves and there are few structures in the thorax and abdomen that do not receive branches from this massive nerve.

The sacral parasympathetic nerves control reflexes required for emptying hollow organs that fill periodically. These nerves arise from S3 and S4 (occasionally extending into S2 and S5), and run as the nervi erigentes to structures in the urinogenital and lower visceral areas. The anatomical arrangement of the parasympathetic nervous system ensures that the discharge of any particular nerve has a relatively local and limited action.

In structure, preganglionic nerves of the parasympathetic nervous system are long with few collaterals. Parasympathetic ganglia are found in or near to the innervated organ and the postganglionic neurones are very short (see Figure 2.2). Parasympathetic ganglia are often microscopically small.

THE DUAL FUNCTION OF THE AUTONOMIC NERVOUS SYSTEM

In general, the parasympathetic and sympathetic nervous systems are considered to have opposite actions but this is not always the case, see Table 2.1 for details.

The sympathetic system has a catabolic action, serving to facilitate all internal processes that aid muscular efficiency. At a certain level of activity, the sympathetic system inhibits processes that interfere with muscular efficiency. A good example is the inhibition of digestive processes in an emergency situation requiring intense sympathetic activity. As you may have experienced, intense

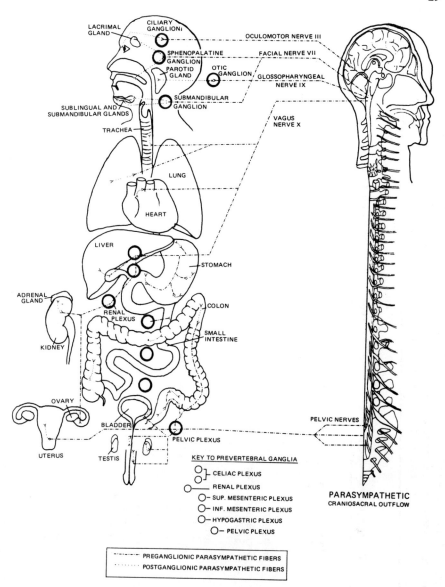

Figure 2.5 A schematic drawing of the parasympathetic nervous system. This figure shows the parasympathetic innervation described in more detail in Figure 2.1.

sympathetic activity following a heavy meal can lead to a leaden and painful sensation in the visceral region due to the inhibition of the digestive processes. The parasympathetic nervous system tends to have an anabolic function and is, in general, concerned with conservation, accumulation, and storing of bodily energies.

Table 2.1 Effects of stimulation on the autonomic nervous system. In general the catabolic or spending action of the sympathetic nervous system is opposed by the anabolic or conserving action of the parasympathetic nervous system. This is not always the case, note reversal of roles for the stomach and pancreas

	Parasympathetic effects	Main receptor type	Sympathetic effects
Eye:			
Iris	Contraction of sphincter pupillae; pupil size decreases	α	Contraction of dilator pupillae; pupil size increases
Ciliary muscle	Contraction; accommodation for near vision	β	Relaxation; accommodation for distant vision
Lacrimal gland	Secretion		Excessive secretion
Salivary glands	Secretion of watery saliva in copious amounts	α	Scanty secretion of mucus – rich saliva
Respiratory system:			
Conducting division	Contraction of smooth muscle; decreased diameters and volumes		Relaxation of smooth muscle; increased diameter and volumes
Respiratory division	Effects same as on conducting division		Effect same as on conducting division
Blood vessels	Constriction		Dilation
Heart:			
Stroke volume	Decreased	β	Increased
Stroke rate	Decreased	β	Increased
Cardiac output and blood pressure	Decreased	β	Increased
Coronary vessels	Constriction		Dilation
Peripheral blood vessels:			
Skeletal muscle	Constriction	α,β	Dilation
Skin	Dilation	α	Constriction
Visceral organs (except heart and lungs)	Dilation	α,β	Constriction

31

Table 2.1 (continued)

	Parasympathetic effects	Main receptor type	Sympathetic effects
Stomach:			
Wall	Increased motility	β	Decreased motility
Sphincters	Inhibited	α	Stimulated
Glands	Secretion stimulated		Secretion inhibited
Intestines:			
Wall	Increased motility	α,β	Decreased motility
Sphincters	Inhibited	α	Stimulated
Pyloric, iliocoecal internal anal	Inhibited		Stimulated
Liver:	Promotes glycogenesis, promotes bile secretion		Promotes glycogenolysis, decreases bile secretion
Pancreas (exocrine and endocrine)	Stimulates secretion		Inhibits secretion
Spleen	Little effect		Contraction and emptying of stored blood into circulation
Adrenal medulla	Little effect		Epinephrine secretion
Urinary bladder	Stimulates wall, inhibits sphincter	α,β	Inhibits wall, stimulates sphincter
Uterus	Little effect		Inhibits motility of nonpregnant organ; stimulates pregnant
Sweat glands	Normal function	α	Stimulates secretion (produces 'cold sweat' when combined with cutaneous vaso-constriction)

Pick (1970) has pointed out that most attempts to provide a comprehensive definition of the autonomic nervous system have failed to allow for the many exceptions and contradictions that are found within the system, and a substantial part of this book is about these differences and contradictions. Most workers have accepted the duality of the autonomic nervous system but it should be pointed out that some have denied a basic and essential dual function (Stöhr, 1949; Meyling, 1953). Stöhr argued that there is no anatomical basis for the concept of antagonistic action between the sympathetic and parasympathetic nervous systems. He considered the entire autonomic nervous system to be a syncytium comprised of interstitial cells having freely anastomosing processes. His idea is reminiscent of Galen's concept of sympathy outlined in Chapter 1.

SMOOTH MUSCLES

Unlike somatic or striate muscles of the body, smooth muscles line the walls of cavities, this means that their contractions are required to be continuous (Huddart and Hunt, 1975). Similarly, smooth muscles around blood vessels require to exert a continuously maintained contraction or tonus. In view of these differences and the fact that, at least at the postganglionic level, autonomic nerve cells are unmyelinated might lead us to expect the electrical discharge of autonomic nervous system nerve cells to show some differences from central nervous system nerve cells. Evidence for such differences comes from two sources. Patton (Ruch *et al.*, 1961) has reported that the electrical discharge, shown by action potentials in autonomic nerve cells, shows slow twitching for periods as long as two seconds. It is argued that this effect is produced by a much slower destruction of the transmitter substance at the nerve junctions of autonomic nerves, as opposed to the rapid breakdown found at the synaptic junctions and end plates of neurones in the central nervous system. Tomita (1970) has indicated that smooth muscles have electrical connections between cells that result in an electonic potential spreading with exponential delay along smooth muscles. Both of these effects would serve to help prolong the contractile process of smooth muscle.

AUTONOMIC TRANSMITTER SUBSTANCES

The enlarged drawing shown in Figure 2.4 shows the relationship between autonomic nerves and their transmitter substances. In common with striated muscles of the central nervous system the transmitter substance at all parasympathetic synapses is acetylcholine (cholinergic). The synapses of the preganglionic nerves in the sympathetic system are also mediated by acetylcholine (cholinergic); however, the postganglionic nerves of the sympathetic nervous system contain noradrenaline which is also the transmitter substance at the postganglionic nerve endings. Sympathetic nerves are traditionally called adrenergic nerves but since von Euler's discovery of the role of noradrenaline in the postganglionic nerves it would be more appropriate to call them *noradrenergic* nerves. The catecholamines stored in the cells of the adrenal medullae are under the control of preganglionic sympathetic nerves called the splanchnics and their release is mediated by the neurotransmitter substance acetylcholine (Hagen, 1959) as is the case with the sweat glands of the body.

The former simple dichotomy between the sympathetic and the parasympathetic nerves proposed by Gaskell and Langley, with its associated pharmacological concept of (nor)adrenergic and cholinergic, is now being shown to be far more complex. The complexity arises from the fact that the earlier histological or electronmicroscopic methods of looking at the two cell types were unable to show essential architectural differences but newer histochemical techniques are able to reveal essential differences. For example, the cell types in the superior cervical gland are 95 per cent (nor)adrenergic and 5 per cent cholinergic and, in addition, in certain species, there appears to be a small number of dopaminergic cells (Bunge *et al.*, 1978). In Chapter 1, Figure 1.2 the sympathetic nerve cells and ganglia were shown as arising from the neural crest cells. These cells migrated out into the body. From embryological tissue-culture experiments it is now argued (Bunge *et al.*, 1978) that cells from the neural crest can give rise to both cholinergic and (nor)adrenergic cells; that is, give rise to both sympathetic and parasympathetic nerves. Moreover, the hypothesis becomes more complex because Bunge and his collaborators argue that the cell typing does not take place until the cell has migrated to its final position. This means that the particular transmitter substance that the cell will eventually use is influenced by the local tissue environment.

To avoid overwhelming the reader with a mass of detail this chapter has concentrated upon the general principles and the reader who requires additional and more detailed information is referred to the many excellent individual chapters in the book by Hubbard (1974).

SENSORY NERVES OF THE AUTONOMIC SYSTEM

Langley considered the autonomic nervous system to be a motor output system without any true sensory input from the peripheral parts of the body. He was aware that the white rami communicantes carried sensory nerves from the viscera, but felt that they should not be considered as proper sympathetic fibres because they lacked a peripheral synapse that was one of the defining characteristics of sympathetic nerves. However, because returning sensory pathways have not been clearly identified, it would be wrong to conclude that the autonomic nervous system does not have a feedback system. Part of the confusion about the nature of sensory nerves appears to arise from surgical and clinical observers attempting to identify autonomic sensory nerves by subjecting the viscera to prods and pricks. This type of stimulation readily elicits feeling in the somatic system, but autonomically innervated structures appear to have different quantitative and qualitative characteristics as well as being more sparsely distributed. Some evidence for autonomic sensory nerves has been claimed by Pick (1970). If these different pathways were confirmed we would have evidence of direct feedback pathways for the autonomic system.

Perhaps the best but little understood example of an autonomic sensory system concerns visceral pain. Visceral pain is said to be severe and intractable and it is common to use surgery in attempts to control it. The initial theories concerning pain held that it was a non-specific function of the number of nerves firing in the brain; an increase in pain was due to an increase in the number of nerve cells firing in the brain. Melzack and Wall (1965) proposed an ingenious model of pain called the gate control theory. Briefly, the theory proposed that pain perception was modified by cells in the dorsal horn of the spinal cord, called the substantia gelatinosa, influencing higher centres in the brain. The model proposed that slow firing sensory nerves, called C fibres, carried neural information concerning pain and that these are modified by fast firing sensory cells, called A fibres, via the cells of the substantia gelatinosa. The cells of the substantia gelatinosa are influenced by nerve inputs at both the local spinal cord level and by the higher centres in the brain. Perception of pain is the net result of the various systems interacting upon the substantia gelatinosa cells. According to the gate control theory, perception of pain may be controlled by the moderating influences upon the cells of the substantia gelatinosa and these may range from the higher cortical influences damping down pain perception or an increase in the firing rate of the fast A fibres. An example of the latter effect would be vigorous rubbing of a painful area which serves to increase the A fibres' influence and damp down the pain.

The autonomic nervous system is a diffuse wide-acting system and any sensory feedback might tend to swamp modifying influences that were reacting upon the cells of the substantia gelatinosa. This results in an overall level of felt pain that is difficult, if not impossible, for the patient to control. A major problem for the theory concerns the phenomenon called referred pain where a patient, suffering from a pain that arises from the viscera, refers the pain to another part of his skin which may be considerably displaced from the true source of the pain. It is the displacement of the pain sensation that is difficult for the gate control theory, because it needs to explain how pain at one level in the spinal cord comes to be felt in a totally different level. At the present time no satisfactory explanation appears to have been proposed. A good discussion of visceral pain can be found in Appenzeller (1970) who also discusses a number of alternative theories but they all fail to provide an adequate overall explanation of the peculiarities of visceral pain. The major problem in the area of sensory feedback and visceral pain is related to the methodological difficulties involved in stimulating the diffuse sensory receptors of the viscera in a reliable and quantitative way.

THE SYMPATHOADRENAL SYSTEM

Apart from the extensive peripheral nervous network the sympathetic system releases the catecholamine hormones, adrenaline and noradrenaline, from the adrenal medullary cells. The medullary cells of the adrenal gland are encapsulated by the cells of the adrenal cortex controlled by the adrenocorticotrophic hormone (ACTH) released from the pituitary gland. In Chapter 1 it was noted that the two parts of the adrenal gland arose from different embryological sources. Both of these systems play vital roles in physiological mechanisms serving homeostasis and it is certainly of interest to question their intimate anatomical roles to see if it is more than coincidence that brings them together. Before we attempt to answer this question, we should consider the role of the catecholamines in bodily function.

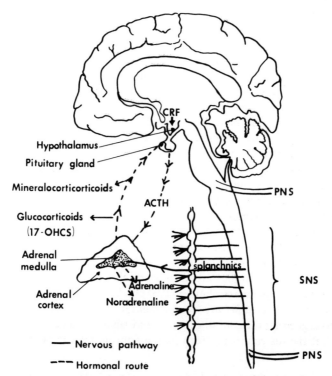

Figure 2.6 The pituitary–adrenal and sympathoadrenal systems. Release of cortical releasing factor (CRF) from the hypothalamus stimulates the production and release of the adrenocorticotrophic hormone (ACTH) from the pituitary gland situated at the base of the brain. ACTH passes into the blood stream and triggers the cells of the adrenal cortex to produce mineralocorticoids and gluco-corticoids that stimulate cells throughout the body and also act as a positive feedback mechanism to reduce the level of ACTH being released from the pituitary gland.

The sympathoadrenal system uses direct nervous innervation for its first stage. Stimulation of the splanchnics causes the release of the catecholamines, adrenaline and noradrenaline from the medullary cells of the adrenal.

Note that the relative reaction speeds of the two systems are different, with the directly innervated sympathoadrenal system being faster.

Adrenal cells containing catecholamines can be stained by certain iron salts and as a result they are called chromaffin cells. It can be seen from Figure 2.6 that the adrenal medullary cells are innervated by preganglionic nerve fibres and this means that the release of adrenal catecholamines is mediated via the neurotransmitter substance acetylcholine. It has been claimed that adrenaline and noradrenaline are manufactured by separate medullary cells (Douglas, 1966).

Adrenaline always represents the largest proportion of the total amount of catecholamine released from the medullary cells, noradrenaline consists of a small fraction of the total amount. Most of the noradrenaline in the peripheral part of the body comes from its release at the ending of the sympathetic neurones (see Figure 4.4). It has been suggested that some of the stored adrenal catecholamines are under the control of the parasympathetic nervous system (Vassalle *et al.*, 1970) but this does not appear to have been confirmed.

In addition to the chromaffin cells of the adrenal medullae there are additional extrachromaffin cells scattered throughout the abdominal region that produce and release catecholamines. The presence of these cells serves to boost the level of circulating catecholamines (Lempinen, 1964). Incidentally, it can also be seen that these extrachromaffin cells might act to confound any study, made to assess the role of catecholamines released from the adrenal medullary cells, that did not take their presence into account.

As pointed out in Chapter 1, Cannon believed that the nervous network of the autonomic nervous system, together with the output of the adrenal gland, represented a single system. But later workers have questioned Cannon's statement, pointing out that the physiological actions of the autonomic nervous network and the catecholamines are sufficiently dissimilar for them to represent two quite separate systems (Celander, 1955). Whatever the outcome of this fine distinction the two systems are complementary and at the present time it is difficult to argue that the overall gain made by treating them as separate systems would offset the losses arising from such a conceptual separation.

PHYSIOLOGICAL ACTION OF ADRENALINE

Adrenaline, with its hormone-like characteristics, has the wider action of the two catecholamines released from the adrenal gland. It produces a marked rise in arterial blood pressure of muscles and the cardiovascular system, and also mobilizes the glycogen reserves stored in bodily tissues. Despite a general action of vasodilation within the internal organs, the peripheral action of adrenaline in the skin is predominantly vasoconstriction. The reason for this is that in an emergency situation, the internal organs and muscles require a large supply of blood, while at the same time, it is clearly of adaptive value in a threatening situation for an animal to have the surface of its skin in a vasoconstricted state to reduce bleeding. An adrenaline swab-stick is an essential item for every boxer's second to help him to facilitate the process of vasoconstriction, should the boxer be cut during the contest. Adrenaline causes a general increase in blood sugar level throughout the body; this is thought to occur via a variety of physiological actions. During an emergency, adrenaline also produces an additional source of energy by releasing fatty acids that circulate in the blood stream. These fatty acids are utilized by the body during the course of the emergency to provide additional energy but, as pointed out earlier, if this energy source is not used, because of social conventions requiring inaction or perhaps the person is sitting relatively inactive in his motor car, then these fatty acids can detrimentally circulate throughout the body for considerable periods of time.

Goffard and Perry (1951) have shown that adrenaline increases the efficiency of skeletal muscles by the paradoxical action of reducing the firing threshold of the muscle and reducing muscular excitement. This muscular reduction is probably related to the relaxation of muscles attempted by highly trained athletes during peak athletic performances.

PHYSIOLOGICAL ACTION OF NORADRENALINE

Most of the circulating noradrenaline in the body comes from its release at the endings of the (nor)adrenergic postganglionic nerves. The action of noradrenaline is not so widespread as adrenaline, but it plays an important role in maintaining cardiovascular and sympathetic tone (Folkow, 1956).

The vascular bed is considerably larger than the vascular volume, and apart from vital areas that require fairly constant levels, blood is constantly being shunted into areas where it is required. For example, following a heavy meal, it is necessary for the nutrients from the digestive system to be carried to the liver for storage; hence, the hepatic portal system is very active at these times and we experience the lethargic feeling of postprandial torpor. Conversely, during a stress or an emergency situation, the essential vascular need is for an adequate supply to the brain and muscles to allow them to cope with the situation. It is one of the primary roles of the sympathetic nervous system and noradrenaline to modify the vascular system to allow blood to be directed to required areas via their action on the contractility of smooth muscles surrounding blood vessels.

During short-term hypothermia both adrenaline and noradrenaline play a role in the responses such as piloerection; also, both of the amines can mobilize fat reserves and cause increased oxygen consumption. Noradrenaline has been shown to play an important role in temperature regulation under conditions of prolonged hypothermia. Jansky (1971) has reported that metabolic changes in response to cold are mediated by noradrenaline. The body temperature effects mediated by noradrenaline are called non-shivering thermogenesis. The specificity of noradrenaline in this temperature role has been shown experimentally by using curarized rats who, under hypothermic conditions, were still able to maintain body temperature as long as their adrenals were intact. Noradrenaline also plays an important role in neonatal thermoregulation in mammals via its action on the 'brown fat' deposits, found in the young (Smith and Horowitz, 1969), which are thought to act as special reserves of energy for low temperature conditions. This temperature regulation function will be discussed at greater length in the next chapter.

THE VASOMOTOR ACTION OF CATECHOLAMINES

One of the most important roles for the sympathetic nervous system is to maintain vascular tone of the venous and arterial blood systems. If we look at the action of the catecholamines in controlling blood supplies, we find many contradictory effects. For example, we have considered the emergency vasoconstrictive effects of adrenaline in the skin and contrasted these with its vasodilation effects within internal organs. These effects are relatively simple to understand compared with vascular effects that can occur within short distances in the same artery or vein. Cannon attempted to resolve this problem by arguing that adrenaline had an inhibitory and an excitatory factor, but his concept, although highlighting the problem, did not solve it. Ahlquist (1948) extended earlier pharmacological concepts by suggesting that the vascular system contained two types of receptors which he called alpha- and beta-receptors. Both types of receptors are stimulated by the sympathetic nervous system but they have opposite actions. Alpha-adrenergic receptors are associated with functions that are inhibited during activation of the sympathetic nervous system, while beta-adrenergic receptors are associated with functions that are stimulated during sympathetic nervous system activity. For example, we would expect to find a high proportion of alpha-adrenergic receptors in the arterioles found in the gut region while beta-adrenergic receptors would form the highest proportion in arterioles supplying blood to muscles. The reason for this is that during sympathetic activity the activities of the gut region tend to be reduced while the activities of muscles are increased. The arterial vascular system contains a mixture of both types of receptors but there are differences in the relative proportions. Alpha-adrenergic receptors predominate in the arterioles of the skin and mucosa while beta-adrenergic receptors predominate in the arterioles of the heart (Schneiderman *et al.*, 1974). In general, veins contain mainly alpha-receptors and, in consequence, the dominant sympathetic effect on the venous vascular system is vasoconstriction. In actual fact the overall picture is, unfortunately, not quite so clear as was initially thought, but the important thing to realize at this stage is that a mixture of the alpha-adrenergic and beta-adrenergic receptors can produce an appropriately graded response in the different parts of the vascular system.

Catecholamines released into the peripheral part of the body via the sympathoadrenal system play an important supportive role for the nervous network of the autonomic nervous system. Cannon argued that the main role for the system was during emergency reactions, but it can now be seen to have a much wider function and plays an important part in the normal homeostatic functions of the body. Cannon's statement limiting the system to an emergency role may have been due to the particular experimental technique he used to examine the system. His technique involved denervating the heart and observing the effects of the catecholamines released from the adrenal gland. To obtain reasonable changes in heart-rate requires fairly massive stimulation of the adrenals, and the technique is not likely to reveal any subtle changes in the heart-rate.

THE PITUITARY-ADRENAL SYSTEM

Cells forming the adrenal cortex are controlled by the adrenocorticotrophic hormone (ACTH) released from the pituitary gland situated at the base of the brain (see Figures 2.6 and 3.5). Release of ACTH causes the cortical cells of the adrenal gland to enlarge and produce vital steroids. Adrenal corticosteroids can be classified into two main groups, mineralocorticoids and glucocorticoids. Glucocorticoids influence a wide variety of biochemical processes within individual cells and have a marked effect on muscular efficiency. The most important mineralocorticoid is aldosterone which plays an important role in water retention by the kidneys. Release of corticosteroids causes the concentration of ACTH to fall. The whole system forms a classic negative feedback loop.

Cannon, emphasizing the role of the sympathetic nervous system during emergency states, largely ignored the pituitary-adrenal cortex system. More importantly, he appears to have overlooked the fact that an emergency might require the overall bodily 'homeostat' to be set to a new level. It was left to Selye (1950) to discover and elucidate the role of the adrenal cortex in stress. Selye's essential findings are encompassed by his term 'general adaptation syndrome'. In many ways Selye's concept is complementary to Cannon's 'flight or fight' concept, but discussion of this point will be left until the final chapter.

Comparison of the sympathoadrenal and pituitary-adrenal systems reveals that steroids have the more general and fundamental bodily role, their major function is maintenance of cellular integrity. Reduction of steroid levels lowers the basal metabolic rate of the body and produces marked muscular weakness. In contrast, the hormonal role of the sympathoadrenal system has as its main function, elicitation of reactions from tissues and organs that are primed by the corticosteroids.

Until fairly recent times, it was assumed that the pituitary-adrenal and the sympathoadrenal systems were largely independent and that the juxtaposition of the adrenal cortical and medullary cells at the apex of the kidneys was largely coincidental. However, recent work has indicated that the two systems are more closely related than was previously suspected. Pohorecky and Rust (1968) have examined the interaction between the two systems and found marked depletions in the catecholamine content of medullary cells of up to 42.5 per cent following removal of the pituitary gland. More recently, Pohorecky and Wurtman (1971) have proposed a theoretical biochemical model involving a relationship between glucocorticoid concentration and the secretion of adrenaline. Clearly in the next few years we will come to have a greater understanding of how these two vital systems interact to maintain normal bodily function.

It is of interest to note that Gellhorn et al. (1941) proposed that insulin was a hormone specifically related to the parasympathetic nervous system. This would mean that the parasympathetic nervous system controlled a hormonal system in a similar way to the control of adrenal catecholamines by the sympathetic nervous system. However, the suggestion of Gellhorn and collaborators does not appear to have been generally accepted.

REFERENCES

Ahlquist, R.P. (1948). 'A study of the adrenergic receptors', *Amer. J. Physiol.*, **153**, 586-600.

Appenzeller, O. (1970). *The Autonomic Nervous System*. North-Holland, Amsterdam.

Blackman, J.G. (1974). 'Function of autonomic ganglia', Chapter 9 in J.I. Hubbard (Ed.) *The Peripheral Nervous System*. Plenum Press, New York.

Bunge, R., Johnson, M. and Ross, C.D. (1978). 'Nature and nurture in development of the autonomic neuron', *Science*, **199**, 1409-15.

Celander, O. (1955). 'The range of control exercised by the sympathico-adrenal system', *Acta Physiologica Scandinavia*, **32**, Supplement 116.

Douglas, W.W. (1966) 'The mechanism of release of catecholamines from the adrenal medulla', *Pharmacol. Rev.*, **18**, 471-80.

Ebbesson, S.O.E. (1968). 'Quantitative studies of superior cervical ganglia in varieties of primates including man. 1. The ratio of preganglionic fibres to ganglionic neurones', *J. Morphol.*, **124**, 117-31.

Folkow, B. (1956). 'The nervous control of the blood vessels', in R.J.S. McDowall (Ed.) *The Control of the Circulation of the Blood*. Dawson, London.

Gellhorn, E., Cortell, L. and Feldman, J. (1941). 'The effect of emotion, sham rage, and hypothalamic stimulation on the vago-insulin system', *Amer. J. Physiol*, **133**, 532-41.

Goffard, M.I. and Perry, W.L.M. (1951). 'The action of adrenaline on the rate of loss of potassium ions from unfatigued striated muscle', *J. Physiol*. **112**, 95-101.

Hagen, P. (1959). 'The storage and release of catecholamines', *Pharmacol. Rev.* **11**, 361-73.

Hubbard, J.I. (1974). *The Peripheral Nervous System*. Plenum Press, New York.

Huddart, H. and Hunt, S. (1975). *'Visceral Muscle: Its Structure and Function'*. Blackie, Glasgow.

Jacobowitz, D.M. (1974). 'The peripheral autonomic system', Chapter 5 in J.I. Hubbard (Ed.) *The Peripheral Nervous System*. Plenum Press, New York.

Jansky, L. (1971). *Nonshivering Thermogenesis*. Sweets and Zeitlinger, Amsterdam.

Lempinen, M. (1964). 'Extra-adrenal chromaffin tissue of the rat and the effect of cortical hormones on it', *Acta Physiologica Scandinavica*, **62**, Supplement 23.

Melzack, R. and Wall, P. D. (1965). 'Pain mechanisms: a new theory', *Science*, **150**, 971-9.

Meyling, H.A. (1953). 'Structure and significance of the peripheral extension of the autonomic nervous system', *J. Comp. Neurol.*, **99**, 495-543.

Pick, J. (1970). *The Autonomic Nervous System: Morphological, Comparative, Clinical and Surgical Aspects*. Lippincott, New York.

Pohorecky, L. and Rust, J.H. (1968). 'Studies on the cortical control of the adrenal medulla in the rat', *J. of Pharmacol. Exp. Therapeutics*, **162**, 227-37.

Pohorecky, L.A. and Wurtman, R.J. (1971). 'Adrenocortical control of epinephrine synthesis', *Pharmacol. Rev.*, **23**, 1-35.

Ruch, T.C., Patton, H.D., Woodbury, J.W. and Towe, A.L. (1961). *Neurophysiology*. Saunders, New York.

Schneiderman, N., Francis, J., Sampson, L.D. and Shwaber, J.S. (1974). 'Central nervous system integration of learned cardiovascular behaviour', in L.V. Di Cara (Ed.) *Limbic and Autonomic Nervous Systems Research*. Plenum Press, New York.

Selye, H. (1950). *Stress: The Physiology and Pathology of Exposure to Stress*. Acta Inc., Montreal.

Smith, R.E. and Hortwitz, B.A. (1969). (Brown fat and thermogenesis', *Physiol. Rev.*, **49**, 330-425.

Stöhr, P., Jnr. (1949). 'Studien zur normalen und pathologischen histólogie vegetative ganglien III, *Zeitschrift fur Anatomie und Entwicklungsgeschichte*, **114**, 14-52.

Tomita, T. (1970. 'Electrical properties of mammalian smooth muscle', E. Bulbring, A. Brading, A. Jones, T. Tomita (Eds.), Smooth Muscle. Arnold, London.

Vassalle, M., Mandel, W.J. and Holder, M.S. (1970). 'Catecholamine stores under vagal control'.

CHAPTER 3

The Autonomic and Central Nervous Systems

It has been argued that not only was the autonomic nervous system autonomous but that it actually controlled brain functions (Kempf, 1920). Most investigators of the autonomic nervous system and its functions must have, at times, felt the appeal of postulating such a dominance but it would be difficult to make out a complete case for such a view. As the functions of the brain and the autonomic nervous system become more clearly understood, the independence of the latter is seen to decrease (Obal, 1966; Germana, 1969). It is the purpose of this chapter to summarize findings about the various parts of the central nervous system that are known to have important roles in autonomic activity.

THE SPINAL CORD

In common with the other vertebrates, mammals are segmental animals. In humans we find that specialization has resulted in numerous displacements but the basic segmental pattern remains and it can be identified in Figure 2.1. By separating the spinal cord at the level of the T1 outflow a purely segmental mammal can be produced. The result of such a separation is catastrophic and all autonomic functions becomes severely depressed. Temperature regulation becomes virtually nil and blood pressure falls to very low levels because the sympathetic nervous system is not able to exert its usual tonus on the blood vessels and the resulting dilation of the blood vessels produces considerable enlargement of the vascular system. After a period of some weeks the depressed autonomic functions slowly reappear but they rarely assume their former smooth and untroublesome action. The severe depression of autonomic function, arising from separation of the spinal cord from the brain, indicates the extent of the influence of higher brain centres.

In common with the somatic motor system the autonomic nervous system has nerve tracts within the spinal cord that run to and from the brain. However, these tracts are neither so distinct nor as clearly demarcated as those of the somatic motor system, and the exact location of many autonomic nervous system tracts has yet to be commonly agreed.

AUTONOMIC FUNCTIONS OF THE HINDBRAIN AND RETICULAR ACTIVATING SYSTEM

The hindbrain contains many specific areas that are crucial for functions of the autonomic nervous system. Comparative studies at the level of the hindbrain of birds and mammals have shown many homologous structures (Ariens-Kappers, *et al.*, 1960). This is not the case for the higher areas of the central nervous system where it is often difficult to identify comparable brain structures. Hindbrain structures are mainly concerned with adaptive reflexes that require the coordination of several different physiological mechanisms. Good examples of this type of reflex are the coughing and vomiting actions (Ranson and Clark, 1953) that require integration of the abdominal muscles together with those of the stomach, diaphragm, and larynx. In these reflexes we see somatic muscles of the central nervous system and smooth muscles of the autonomic nervous system in coordinated synchrony.

One of the most important structures within the hind and lower midbrain is the reticular activating system which has been accorded a major role in numerous theories of behaviour. Up to the present time most attempts have tried to relate it to the higher brain functions; in particular, to seek for the mechanisms by which the reticular activating system controlled the arousal level of the cerebral cortex.

In terms of evolution it is probable that, at a certain period of time in nervous tissue development, an animal existed having an internal vegetative nervous system and a brain that was little more than a hindbrain with a general activating role. This system would roughly correspond to what we now understand as the reticular activating system. Thus, we might expect to find a special and close relationship between the reticular activating system and the autonomic nervous system arising from an earlier evolutionary stage. This statement is speculative but there a few clues in the literature. Bonvallet *et al.* (1954) and Dell *et al.* (1954) have suggested a feedback circuit for the autonomic nervous system that is located in the reticular activating nervous system. The feedback circuit takes the form of adrenaline-sensitive cells located in the reticular activating system that detect blood-born adrenaline released from the adrenal glands. The function of these adrenaline-sensitive cells is to modulate the level of the reticular activating system and thence cortical activity. These authors state that increasing levels of adrenaline in the body correlate positively with cortical activity. It has been claimed that a similar correlation exists between sympathetic tone and electrical activity of the cortex. In addition, Rothballer (1956) has reported a similar finding between levels of noradrenaline and cortical activity. More recently Baust and Niemezyk (1963) have reported that they were unable to confirm these earlier findings despite using high levels of injected adrenaline. However, these findings are usually reported as part of a general overall phenomenon of arousal and it is rare for any detailed analysis to be made and the exact details have yet to be worked out.

AUTONOMIC REFLEXES OF THE HINDBRAIN

Electrical, chemical, and surgical techniques have revealed groups of cells located in the reticular formation, areas of the pons and the medulla oblongata (see Figure 3.1) whose major functions are vasomotor, cardiac, and respiratory reflexes. Discrete groups of cells are found to be concerned with the processes of swallowing and intestinal movements of the digestive processes. Other groups of cells control the processes of micturition. Smaller groups of nerve cells are concerned with salivation and control of the size of the pupil in the eye. The subtle interactions between the somatic and the autonomic nervous systems are beautifully illustrated by the swallowing reflex. When a bolus of food leaves the mouth it is initially under the control of the somatic nervous system but midway down the oesophagus the peristaltic movement of the food changes from control by the somatic nervous system to control by the autonomic nervous system. In general, the basic circuits controlling the autonomic processes form servo-feedback loops which can be modified by higher cognitive or emotional centres in mid and forebrain structures but the basic reflexes cannot, usually, be completely suppressed.

Respiration is a relatively simple example showing integration between the somatic and autonomic systems. Centres have been identified in the reticular activating system that have been shown to play important roles in the respiratory process. Electrical stimulation of a group of cells in the medulla oblongata produces intake of air, while another group of cells at a lower level in the hindbrain can be shown to produce exhalation of air from the lungs. Air intake or inspiration appears to be the dominating reflex because exhalation of air from the lungs is largely controlled by the elasticity of the lungs following their expansion after taking a breath. Interoceptive receptors situated in each lung transmit nerve impulses to respiratory centres in the hindbrain and, when the impulses reach a certain frequency, intake of air slows down and stops allowing air to be exhaled from the lungs. As mentioned above, exhalation is controlled largely by the elasticity of the lungs but somatic muscles in the diaphragm play a part in this process. As pressure inside the lungs falls nervous impulses begin to increase and, at a certain frequency, the air inspiration centre in the medulla oblongata causes the air intake phase of the cycle to begin again. In addition to control by basic reflex loops, the respiration centres are sensitive to gas gradients within the blood that is passing through them. For example, an increase in carbon dioxide levels in the blood stream causes an increase in the rate and depth of respiration which in turn produces a reduction in the carbon dioxide level in the blood by increasing the rate of its removal. In addition to the controls mentioned above, respiratory centres are modified by interoceptive receptors, the carotid bodies, situated in the carotid arteries that react to changes in gas gradients of the blood.

BLOOD PRESSURE

In order for blood to circulate around the body a pressure gradient is required and the basic reflexes for this gradient demonstrate an example of the integration between the sympathetic, parasympathetic, and the somatic nervous systems, as well as the neuroendocrine systems of the body. The pressure gradient for a normal person lies between the values of 80 mm, when the heart is momentarily paused, to 120 mm of mercury at the moment when the heart reaches the maximum force of its contraction. The difference between these two values, measured by a device called a sphygmomanometer, represents the efficiency of the heart. Mechanoreceptors in the carotid arteries and the walls of the aortic artery transmit nervous information to cardiac and vasomotor centres in the medulla oblongata via vagal (X, cranial nerve) and glossopharyngeal (IX, cranial nerve) efferents (Gardner, 1975). Mechanoreceptors are stretch receptors in the arterial walls and they are sometimes called baroreceptors. Major groups are found in the carotid sinus and the aortic arch.

A schematic diagram showing the basic connections is given in Figure 3.1. Nerve fibres from the cardiac centres pass to the heart where they produce a reflex slowing of heartrate. Nerve fibres from the vasomotor centre descend to the spinal cord where their combined neural information is integrated to produce vasodilation. As indicated in the last chapter sympathetic nerves

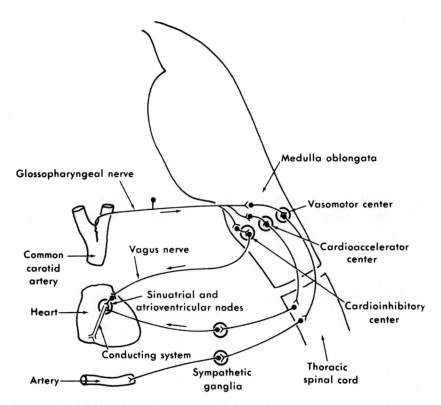

Figure 3.1   Hindbrain structures controlling blood pressure. Nerve fibres from the cardiac centres in the hindbrain pass out to reach the heart where they produce a reflex slowing of the heart-rate. Feedback information is relayed back to these centres from the carotid bodies and the arterioles. Reproduced, with permission, from Gardner, *Fundamentals of Neurology*, W.B. Saunders Company, 1975.

control the diameter of the peripheral arterioles. An increase in the diameters of the arterioles results in a fall of blood pressure which, in turn, results in a decrease in the frequency of the impulses back to the hindbrain from interoceptive receptors located in the carotid bodies and the aortic artery. Under resting conditions this basic reflex produces slight fluctuations in blood pressure. In addition to the basic circuits mentioned above there are other pressure-sensitive cells in the heart. The basic circuits dealt with so far are known as intrinsic neural adjustment mechanisms for blood pressure but other factors are also important, these are called extrinsic factors. Hormones play important roles in the control of blood pressure and the vital roles of the catecholamines were mentioned in the last chapter. In addition, under control of the autonomic nervous system and hormones (Gardner, 1975), rennin is released into the circulation where it leads to the generation of a hormone called angiotensin which plays an important part in the control of fluid levels in the body by releasing the hormone aldosterone, an electrolyte-regulating steroid, from the adrenal cortex. Angiotensin is also a potent dipsogen acting via the septum and the anterior part of the hypothalamus (see Figure 3.6). Schneiderman *et al.* (1974) have said that although the bulbar reticular formation is clearly critical in the regulation of the cardiovascular system, blood pressure involves neural integration at virtually every level of the central nervous system from the brain to the spinal cord. The influence of the anterior hypothalamus and the preoptic region (see Figure 3.6) provides a good example of higher brain involvement because cooling the blood circulating through these areas of the brain results in the arterioles of the skin vasodilating. Electrical stimulation of the anterior hypothalamus results in a fall in arterial blood pressure and increased heartrate. Lesion studies involving cardiovascular responses suggest that an intact cortex is required (Schneiderman *et al.*, 1974). Studies involving extrinsic influences upon blood pressure clearly indicate the importance of higher brain areas and illustrate that in normal cardiovascular adjustments we find a constellation of behavioural and autonomic responses that involve many areas of the central nervous system.

BIOGENIC AMINES

During the 1950s acetylcholine was the neurone transmitter substance about which most was known. Many investigations were carried out as to its role as the chemical mediator at the synapses of the somatic nervous system and preganglionic terminals of the autonomic nervous system as well as its function in the central nervous system. The role of acetylcholine as a neurotransmitter substance and its effects on behaviour have been comprehensively discussed in a book by Warburton (1975). This chapter will only discuss acetylcholine where it has been shown to have direct effects on autonomic functions, but interactions between different neurotransmitter substances are very important and they should not be overlooked. Following the discovery that catecholamines could be made to fluoresce, great interest in fluorometric techniques developed and this has resulted in a considerable body of knowledge being collected about the role and action of adrenaline and noradrenaline (Udenfriend, 1962; Iversen, 1967, 1973, and 1975; Nagatsu, 1973; Blaschko and Muscholl, 1972; Usdin and Snyder, 1973). Fluorescence occurs when a substance has the property of absorbing radiation at one wavelength and emitting it at another. The phenomenon can be seen in many lubricating oils. These oils are usually brown in colour; however, when held up and illuminated by sunlight they fluoresce and become greenish or bluish in colour. Not only have the various fluids of the body been analysed for adrenaline and noradrenaline content, but histochemical techniques (formaldehyde vapour is allowed to condense on thin sections of cells and neurones) enable the presence of adrenaline or noradrenaline to be detected in various tissues. The present knowledge about the role and action of adrenaline and noradrenaline makes, by comparison, the knowledge about acetylcholine look relatively meagre and thin.

Apart from adrenaline and noradrenaline there is one other catecholamine, dopamine, that is thought to have a role as a transmitter substance. The metabolic steps in the biosynthesis of these three catecholamines are shown in Figure 3.2 which shows the catechol nucleus and the ascending steps that are mediated by various enzymes. The actual metabolic steps are from the amino acid, phenylalanine to tyrosine-dopa-dopamine-noradrenaline-adrenaline. Noradrenaline and adrenaline are found in peripheral parts of the body; noradrenaline and dopamine mediate important (nor)adrenergic systems in the brain. Adrenaline is not present in any considerable amounts in the mammalian brain although it has been reported as present in the amphibian brain (Vogt, 1959, 1973).

Catecholamines are pharmacologically active in the brain but peripheral levels do not affect the brain as dopamine and noradrenaline do not cross the blood-brain barrier in any quantity. The blood-brain barrier is thought to operate via the astrocyte cells surrounding the neurones providing a dynamic two-way path between the nerve cells and the capillaries of the cerebral vascular system.

Vogt (1973) has pointed out that despite the widespread nature of the (nor)adrenergic system in the brain, it can sustain extensive damage and disruption without causing very conspicuous signs. The exception to this state-ment is the dopamine system consisting of the nigro-strial tract; damage to this tract results in severe motor disability in man. Dopamine is thought to be a transmitter substance for the important extrapyramidal system of sensorimotor behaviour. This particular sensorimotor system consists of a complex series of loops that ramify into most areas of the brain but has as its main centre, the basal ganglia that lie immediately below the cerebral cortex. Damage to this system results in Parkinson's disease in which patients suffer from unintentional tremor when their limbs are at rest. Treatment consists of giving patients doses of L-dopa, which unlike dopamine can cross the blood-brain barrier, to enable them to build up their levels of the transmitter substance dopamine (Hornykiewicz, 1973). Treatment with L-dopa has been found to be successful in a large number

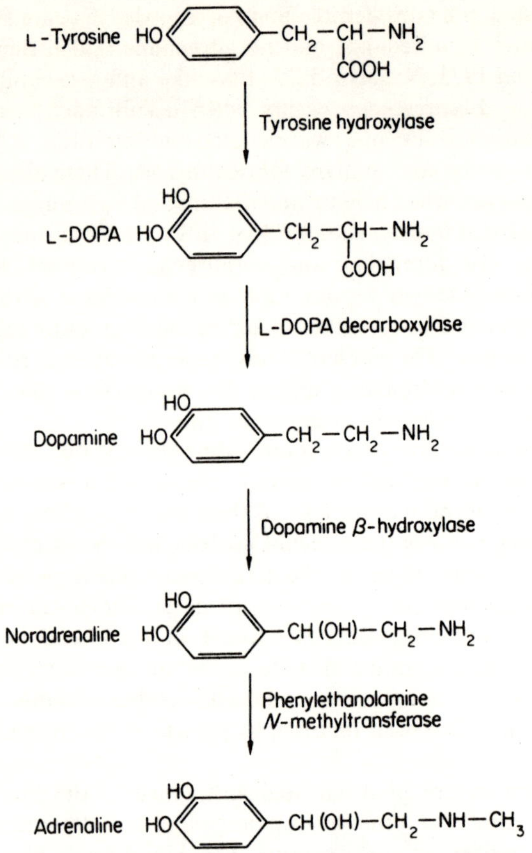

L-Tyrosine HO—⟨benzene ring⟩—CH$_2$—CH—NH$_2$
                                        |
                                       COOH

↓ Tyrosine hydroxylase

L-DOPA HO / HO—⟨benzene ring⟩—CH$_2$—CH—NH$_2$
                                      |
                                     COOH

↓ L-DOPA decarboxylase

Dopamine HO / HO—⟨benzene ring⟩—CH$_2$—CH$_2$—NH$_2$

↓ Dopamine β-hydroxylase

Noradrenaline HO / HO—⟨benzene ring⟩—CH(OH)—CH$_2$—NH$_2$

↓ Phenylethanolamine
   N-methyltransferase

Adrenaline HO / HO—⟨benzene ring⟩—CH(OH)—CH$_2$—NH—CH$_3$

Figure 3.2 The biosynthetic pathway of dopamine, noradrenaline, and adrenaline. The figure shows the biochemical transformations made to the catechol nucleus by enzymes (tyrosine hydroxylase, L-dopa decarboxylase etc.) that mediate the biosynthesis steps needed to form the three active catecholamines, dopamine, noradrenaline, and adrenaline. Although dopamine and noradrenaline are precursors of adrenaline they are important neurotransmitter substances in their own right. Reproduced, with permission, from Blaschko, H. (1973) 'Catecholamine Biosynthesis'. In: *Br. Med. Bull.*, Vol. 29, No. 2, pp. 105–109. Figure 1.

of patients. For the purposes of this book, it is worth pointing out that dopamine provides another example of a direct link between the autonomic and somatic nervous systems. It is obvious that if the position of the arm is altered, the peripheral sympathetic nervous system must make adjustments within the vascular system to provide blood to the new position of the arm. This role of vascular monitoring is a continuous process, albeit a less obvious one than the more dramatic role of the sympathetic nervous system during emotion and stress but it is important to remember that the difference is one of degree. The sympathetic nervous system does not only have the role of 'switching in' during emergency states.

5-Hydroxytryptamine (serotonin) is also an important biogenic amine that is found in the brain in large amounts. Its action is very similar to the catecholamines but it is an indole amine rather than a catechol amine. During the period of the 1960s it was proposed that serotonin played a major role in mood and feelings (Wooley, 1962, 1967); in addition, it is important for sleep processes (Jouvet, 1969). Serotonin is not now thought to be a transmitter substance but to act more like a hormone with an important role in controlling the cerebral vascular system. This is interesting as substances traditionally considered as hormones were thought to be excreted from discrete endocrine glands. More recently Horovitz *et al.* (1972) have suggested that noradrenaline has a hormonal role in which it induces enzymatic activity in an enzyme in the cellular wall which, in turn, produces a messenger that enters the main body of the cell and alters the metabolic processes of the cell. The explanation of these hormone-like roles probably lies in the fact that apart from their known roles as transmitter substances the biogenic amines have additional effects on the numerous brain cells that are not nerve cells. Recent developments in histochemical techniques have enabled the (nor)adrenergic pathways and the serotoninergic pathways of the brain to be traced (Understedt, 1971). From these studies it has been shown that the hypothalamus plays an important role in the neurochemical pathways that ascend and descend to hindbrain centres. The main routes of some of these pathways have been traced, see Figures 3.3, 3.4 and

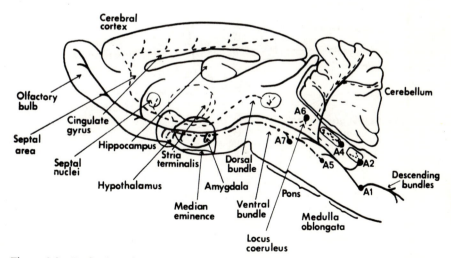

Figure 3.3 Sagittal section of the rat brain showing the principal ascending and descending noradrenergic pathways. After Ungerstedt (1971). Reproduced by permission of the British Medical Journal.

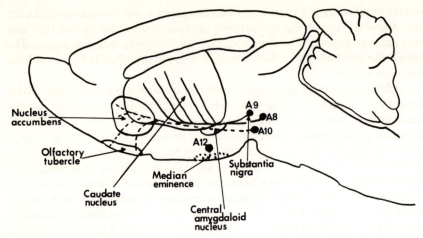

Figure 3.4 Sagittal section of the rat brain showing the principal dopamine pathways. After Ungerstedt (1971). Reproduced by permission of the British Medical Journal.

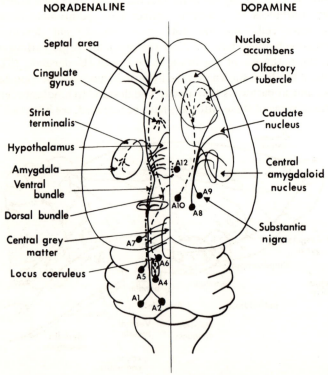

Figure 3.5 Horizontal section of the rat brain showing ascending noradrenergic and dopamine pathways. After Ungerstedt (1971). Reproduced by permission of the British Medical Journal.

3.5. One of the important activities of the hypothalamus is to link the hindbrain and the limbic system, this in part appears to be mediated by the biogenic amines; the hypothalamus is rich in amounts of noradrenaline and serotonin.

In addition, Livett (1973) has described histochemical findings that indicate direct (nor)adrenergic links between the cerebral cortex and the hypothalamus via connections with the ascending and descending pathways between the brain stem and the cortex. Anlezark *et al.* (1973) have shown that bilateral lesioning of one of these clusters of noradrenergic cells, called the locus coerulus, results in the depletion of noradrenaline in the cerebral cortex of the rat. Moreover, these authors argue that learning is specifically affected following ablation of this nucleus. Hartman *et al.* (1972), from the finding that a large proportion of (nor)adrenergic neurones in the brain terminate around blood vessels, have suggested that the locus coerulus might be regarded as a central extension of the sympathetic chain of ganglia. This work is only at the initial developmental stages and a lot remains to be done but it is clearly an area of research that will reveal increasing knowledge about the interaction between the brain and the autonomic nervous system.

THE HYPOTHALAMUS

The hypothalamus consists of complex groupings of nerve cells that lie at the centre of a mass of major incoming and outgoing pathways and connect to most areas of the brain, see Figure 3.6. It is common to divide the hypothalamus into four main regions, these are: anterior, posterior, lateral, and dorsal; these regions are indicated by boxed labels in Figure 3.6. Using electrical and chemical stimulation techniques groups of cells have been identified in the hypothalamus as having a discrete function, these include the following: preoptic, supraoptic, periventricular, ventromedial, dorsomedial, and the mamillary bodies. Since Le Gros Clark's (1938) findings showing connections between the pituitary gland and the periventricular, supraoptic, and tuberal nuclei it has been usual to include the pituitary as part of the hypothalamus.

Following the demonstration by Bernard (1877) of increased levels of blood sugar when the floor of the fourth ventricle was punctured, Karplus and Kreidl (1909, 1910, 1912) showed that autonomic nervous system effects could be obtained by electrically stimulating various points in the midbrain. They found that areas where reliable effects of electrical stimulation could be obtained were shown to involve the hypothalamus which consists of brain structures in the wall and floor of the third ventricle. In humans the hypothalamus represents only one-eightieth of the total brain weight but it is richly supplied with blood vessels, being a complex area containing many important centres of autonomic function. Over the years, its role appears to have been consistently overstated by many early workers who claimed that the hypothalamus controlled the autonomic nervous system (see Symposia: Le Gros Clark *et al.*, 1938; Fulton *et al.*, 1940; Gellhorn, 1943). The most recent overview is contained in Martini *et*

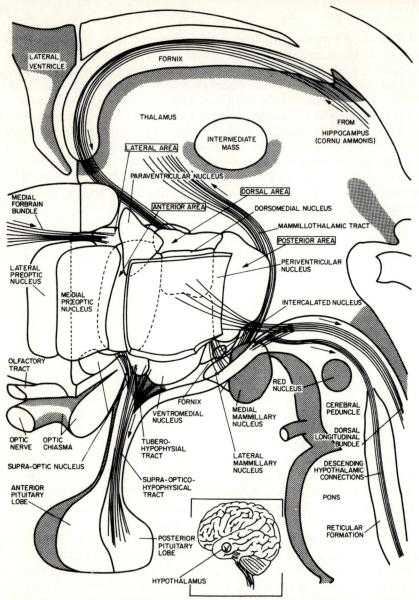

Figure 3.6   The hypothalamus. This diagram shows the human hypothalamus in median sagittal section. The position of the hypothalamus in the brain is shown in the small inset at the bottom of the diagram. The hypothalamus consists of a number of groups of nuclei having marked effects on the autonomic nervous system that lies at the centre of a number of important incoming and outgoing brain pathways. The labels shown in boxes refer to the four major areas: anterior, posterior, dorsal, and lateral. Within the four areas are various groups of nuclei that include: the preoptic, periventricular, ventromedial, and mammillarly bodies. Stimulation of nuclei in the anterior areas of the hypothalamus produces predominant parasympathetic effects in the body. Stimulation of nuclei in posterior areas of the hypothalamus produces predominant sympathetic effects in the body. Reproduced by permission of CIBA Laboratories Ltd.

al. (1970), which is a report of a workshop conference relating to the hypothalamus and its mechanisms.

W.R. HESS

During the course of a long career W.R. Hess studied the relationship between the hypothalamus and the autonomic nervous system and he contributed many of the concepts that are now used to explain the function of the autonomic nervous system (Hess, 1954). His technique was to implant electrodes into thalamic and hypothalamic areas of cats' brains and to study the effects of stimulating these electrodes. It has been claimed that he studied 380 cats and a total of 3,500 stimulation points (Gloor, 1954). His main interest was in the interplay of brain functions (ein kraftespiel) and he showed that stimulation of nuclei in the posterior areas of the hypothalamus produced sympathetic nervous system effects, which he called ergotropic. In the more anterior areas, stimulation produced parasympathetic nervous system effects, which he called trophotropic. Although all areas tended to produce considerable overlap between ergotropic and trophotropic responses and often somatic motor effects were obtained, Hess, nevertheless, concluded that the effects produced in the different clusters of nuclei justified a dual classification into ergotropic and trophotropic zones.

An example of Hess's technique would be that under suitable conditions, the eye could be used to map ergotropic (pupil dilation) and trophotropic (pupil contraction) areas in the hypothalamus. Other hypothalamic areas were shown to be concerned with less specific function such as the sleeping-waking continuum. The idea of the hypothalamus containing a dual organization led to the concept of 'autonomic tuning', Gellhorn (1957, 1967). 'Autonomic tuning' refers to Gellhorn's concept that the brain selects an operating wavelength along the sympathetic-parasympathetic continuum in a similar manner that one would select a wavelength on a radio receiver set. Hess preferred to use a concept he called 'vegetative proprioceptivity', in which he inferred that both sympathetic and parasympathetic effects were tuned and the resulting response depended upon the relative state of readiness of any particular organ.

MOTIVATIONAL CENTRES OF THE HYPOTHALAMUS

Following the failure of early workers (Cannon and Washburn, 1912; Cannon, 1934) to satisfactorily explain hunger and thirst by local effects arising from the cells in the mouth or stomach, attempts were made to seek specific hypothalamic centres controlling these behaviours. A number of models have been proposed but basically they are similar in that they postulate excitatory and inhibitory centres and, in this respect, resemble Hess's hypothalamic model of autonomic function. A good example is Stellar's (1954) model that is reproduced in Figure 3.7 from a redrawing produced in Morgan (1965). This model, which was primarily concerned with the role of the hypothalamus,

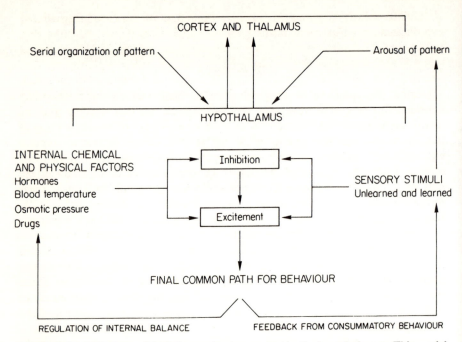

Figure 3.7   Stellar's model of the motivation centres in the hypothalamus. This model involving the hypothalamus gives a general scheme of physiological factors that contribute to the control of motivated behaviour. Note that excitatory and inhibitory centres are prominent in the model. From Physiological Psychology by Morgan. Copyright © 1965 The McGraw-Hill Book Company.

reinstated the importance of social and conditioning factors in motivational behaviour. These important points will be discussed in a later chapter of this book. Hypothalamic control of feeding has been extensively investigated by stimulating or destroying critical areas within the hypothalamus of experimental animals. These studies resulted in a model in which food intake was said to be regulated by a feeding centre in the ventrolateral area of the hypothalamus that was, in turn, under the control of a satiety centre, controlling food intake, situated in the ventromedial area. This model has subsequently been shown to be oversimplified. Reynolds (1965) presented evidence that the method of lesioning can bias the results. Bindra (1969) demonstrated that the earlier view of motivation, being derived from purely internal physiological mechanisms, is too simplistic by demonstrating the importance of external environmental factors as well as the effects on motivation arising from conditioned or learned responses. Grossman (1973) has pointed out that motivational and food-seeking behaviour can be elicited from a number of brain areas outside of the hypothalamus. Finally it must be remembered that the hypothalamus of the rat, the laboratory animal usually being used in this type of study, is very large relative to its total brain size.

One interesting problem is the sensory effects that are found to be associated with lesions or stimulation of the hypothalamus. Hess (1954) indicated that hypothalamic damage in cats often produced sensory defects. MacDonnell and Flynn (1966) explored one aspect of this sensory defect by showing that electrical stimulation of hypothalamic areas in cats could produce violent attack behaviour. Marshall and his collaborators (Marshall *et al.*, 1971; Marshall and Teitelbaum, 1974) have reported a series of experiments that were designed to reveal the nature of sensory defects induced in rats by hypothalamic lesions of the lateral nuclei. Marshall showed that lateral lesions produced marked sensory impairment and, in addition, this impairment could be confined to one or other side of the body. This effect was shown in an experiment where it was found that rats with bilateral lesions confined to one side of the brain showed sensory motor defects on the contralateral or opposite side of the body. Teitelbaum and Epstein (1962) showed that the aphagia and adipsia that are initially found when the lateral hypothalamus is lesioned in rats attenuate over a period of a week or more if the animal is force-fed or given special preferred foods. Marshall has reported that the sensorimotor defects show recovery rates that parallel the recovery from the aphagia and adipsia. In a later paper Marshall (1975) showed that lesions in the hypothalamic medial area result in increased responsiveness of the sensorimotor systems. Bilateral lesions of the ventral area resulted in the increased responsiveness on the contralateral side of the body. By training rats to lever press for food reinforcement when they were stimulated in lateral hypothalamic regions, Beagley and Holley (1977) reported that they were able to show separation of the sensory and motor components. By using bilateral stimulation the authors showed that facilitation was confined to the side being stimulated by pinpoint light stimulation to one or other of the eyes. In addition, the authors showed that the effect reported above is not confined to reflex behaviour but that it can also affect a learned response.

For a more detailed account of the motivational and behavioural centres in the hypothalamus the reader is referred to Grossman (1973) or Buck (1976) if they prefer a more human oriented account. For the purpose of this book it is proposed to concentrate upon the example of bodily temperature control as this was the example given in Chapter 1 as being of great evolutionary importance for mammals. However, before we examine temperature regulation we must first consider the pineal and pituitary endocrine systems and their many interactions with the hypothalamus.

THE PINEAL GLAND

The pineal has a colourful history that ranges from Descartes's 'seat of the soul' to being called a vestigial gland, 'the appendix of the brain'. Its actual role, although neither of these two extremes, is no less fascinating.

In fishes, amphibia, and some reptiles the pineal is a photoreceptor organ containing cells that resemble retinal cells (Kappers; Kappers and Schade, 1965). In mammals the pineal gland appears to have lost its photoreceptive

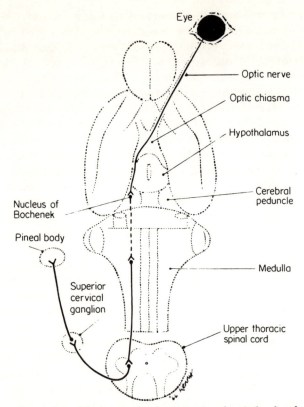

Figure 3.8   The interaction between the pineal gland and
the sympathetic nervous system. Schematic diagram of the
pathway taken by nerves, transducing light impulses,
between the retina and the pineal gland. The pathway is
believed to pass down to the cervical region of the spinal
cord to reach the superior cervical gland of the
sympathetic nervous system. It then passes back up to the
pineal gland in the nervi conarii. After Wurtman *et al.*
(1968), *The Pineal*. Reproduced by permission of
Academic Press, New York.

organelles but retained its light sensing function in that it secretes hormones in
response to environmental light. Wurtman *et al.* (1968) have described the
pineal gland as 'a biological clock'. The importance of the gland for the purposes
of this book is that it receives direct innervation from the autonomic nervous
system via nerves from the cervical ganglia. The nervous connection to the brain
from the autonomic system, so long sought by mediaeval anatomists, has now
been shown to exist. Surprisingly, it turns out to be afferent and not efferent
brain pathway relaying information to the pineal gland via the sympathetic
nervous system. The exact system is conceived of as a feedback loop as follows:

visible radiation received by the eye is transduced into electrical energy where it then passes back into the visual areas of the brain. One major effect in these visual processing centres is that information is passed via various hind and mid-brain structures to the upper thoracic spinal cord where it passes out with sympathetic preganglionic outflows to enter the cervical ganglia. Certain post-ganglionic nerves from the cervical ganglia (nervi conarrii) pass back to the pineal where they end in a dense network on the pineal gland (see Figure 3.8). It has been claimed that the pineal gland in rabbits receive nerves from the para-sympathetic nervous system (Romijn, 1973), but this has yet to be confirmed.

The pineal gland contains large amounts of noradrenaline, serotonin, and the indole amine, melatonin. It is these three chemicals that are believed to control the cyclic activity of the body. Melatonin shows a circadian rhythm with falling levels during the day and increasing levels during the night. Serotonin also shows marked circadian rhythms that can be inverted by reversing the light / dark cycle in experimental animals. Noradrenaline content of the gland shows a threefold variation over a 24- hour period that can be abolished by keeping an animal in constant darkness. These findings seem to suggest that the brain has developed a system for monitoring light which produces cyclic integration of the whole body. The pineal gland serves as the 'master clock' in this process. Other glands exhibit a series of smaller cycles at various points around the 'master clock's' main cycle (Moore *et al.*, 1968; Schapiro and Salas, 1971). Kappers *et al.* (1974) have discussed the role of the pineal gland and its influence on the hypo-thalamus. They felt on balance that a case could not be made at the present time for an endocrine function for the pineal gland on the grounds that specific target organs had not been identified. However, they claimed to have demonstrated an influence by the pineal on the neurosecretory cells of the hypothalamus which in turn controls the sexual cycles of animals.

THE PITUITARY GLAND

Consideration of interactions between the autonomic and the endocrine systems provides an excellent demonstration of the dynamism of living organisms. The hypothalamus is usually considered to be the major centre in the brain for the reflex control of hormones released by the pituitary gland. Situated below the hypothalamus at the base of the brain, the pituitary is often called the master gland of the brain. In view of the more recent ideas mentioned in the last section about the role of the pineal gland, perhaps one should be cautious in giving the pituitary gland this title but in terms of total number of vital hormones secreted it certainly takes the first place.

The pituitary gland is divided into two distinct lobes called the anterior (adenohypophysis) and the posterior (neurohypophysis). The anterior lobe controls the release of five or six hormones and the posterior lobe controls the release of two hormones. Since Scharrer and Scharrer (1963) it has been realized that hypothalamic control of the pituitary gland is via neurochemical pathways rather than by conduction of electrochemical impulses along a nerve cell. The actual mechanisms will be considered in greater detail after we have considered the types of hormones released. It is not proposed to deal with the hormones in any great detail but their important motivational aspects should not be over-looked. For example, a brief consideration of the role of sex hormones *vis-à-vis* behaviour will demonstrate this important point. Additional information will be found in most physiological or psychophysiological textbooks. A good initial source is Gardner (1975) while Leshner (1978) has written a book that specifically relates to the interaction between hormones and behaviour.

The posterior pituitary secretes two hormones vasopressin and oxytocin which between them have four overlapping actions. You may recall that vasopressin was mentioned for its role in controlling blood pressure levels and its action in stimulating water reabsorption from the urine. This antidiurectic action is important and can be shown to be controlled by the hypothalamus, because dehydration of hypothalamic cells results in the release of vasopressin. Oxytocin plays important roles in the contraction of the uterus, part of this action involves stimulation of the smooth muscles of the uterus, it also plays a role in milk production by the mammary glands.

The anterior pituitary lobe plays a part in the regulation of the adrenal cortex, the thyroid gland, and the sex glands. The specific hormones are: somatotropic hormone (STH) which has as it's main action stimulation of normal growth over the whole body; timing and the extent of bodily growth is governed by STH. Thyrotrophic hormone (TTH) acts directly on the thyroid gland causing the release of thyroxine. The next three hormones affect the sex glands of the body. The follicle stimulating hormone (FSH) stimulates development of sperm cells in the male and development of ova in the female. The luteinizing hormone (LH) stimulates the output of sex hormones by causing release of oestrogen from the corpus luteum of females and androgens from the interstitial gonadal tissue of males. The lactogenic hormone (prolactin) plays an important role in the sexual cycles of female sex organs.

The final hormone is the adrenocorticotrophic hormone (ACTH) which stimulates and causes the adrenal cortical cells to release the vital mineralo-corticoids and glucocorticoids mentioned in the last chapter (see Figure 2.6). The importance of this particular hormone and its interaction with the sympathetic nervous system was outline in the last chapter. When considering sympathoadrenal and pituitary-adrenal axes it is important to note that the release of ACTH and stimulation of the anterior lobe of the pituitary gland produces an adrenomedullary discharge of the catecholamines from adrenal medullary cells (Smelik, 1970). The importance of the hypothalamus in controlling the release of ACTH is indicated in Figure 2.6 where the corticoid releasing factor (CRF) is indicated as affecting the anterior lobe of the pituitary gland. The hypothalamus and the pituitary gland are rich vascular areas that are very sensitive to changes in blood concentration or temperature. In addition there are the many hypothalamic neurotransmitter substances and local hormones that interact together in ways that are not fully understood. However, part of the story about the hypothalamic control and release of hormones from the pituitary gland is known. The reader will by now be familiar with the general statement (made from findings using electrical or chemical stimulation of the hypothalamus) that the anterior areas tend to give a parasympathetic response pattern and the posterior areas a sympathetic response pattern. These findings give the view of a brain as consisting of areas that interact. Studies of the hypothalamic cells that innervate the pituitary gland reveal a similar pattern of control by separate areas. Cells in the anterior area of the hypothalamus control the posterior lobe of the pituitary, with the anterior lobe of the pituitary controlled by cells in the posterior part of the hypothalamus. However, there does not appear to be any simple relationship between the hormones and their releasing hypothalamic area in terms of dominating parasympathetic or sympathetic patterns.

Nerve cells in the paraventricular and supraoptic areas of the hypothalamus have neurones that project down into the posterior lobe of the pituitary. These neurones synthesize the hormones oxytocin and vasopressin which are stored in the lobe until released into the general circulation of the body. Release of hormones from the anterior lobe is more complex. The anterior lobe is divided into separate layers that contain different kinds of cells relating to the different hormones. Release of the major pituitary hormones appears to be by local hypothalamic hormones called releasing factors (RF). A number of these releasing factors have been identified and associated with each one as an inhibiting factor. Nerve cells in the tuberohypophyseal region (see Figure 3.6) of the hypothalamus secrete various releasing factors and transport them to a part of the pituitary gland stalk called the medial eminence, which contains a rich vascular system called the primary capillary plexus. Here the releasing factors enter the hypophyseal portal vein to be transported, in a manner that is not fully understood, to appropriate cells in the different layers. Stimulation of these cells causes them to release their hormones into the general circulation of the body where they eventually reach the target gland or cells. Three basic hypothalamic-pituitary feedback mechanisms have been noted by Motta *et al.* (1975). The long feedback mechanism is where the messenger chemical for the hypothalamus comes from hormones synthesized in a hormone gland of the body; an example of this would be the feedback loop of ACTH and the corticoids shown in Figure 2.6. Short feedback mechanisms are where the signals for the hypothalamus are provided by the release of hormones from the pituitary gland. Ultrashort feedback mechanisms are where the signals governing the release of hormones are generated and controlled inside the hypothalamus by the releasing factors themselves. As mentioned earlier a number of releasing factors and corresponding inhibiting factors have been identified (Mess *et al.*, 1970), these are thought to control the release and amount of hormones secreted from the pituitary gland.

In addition to the hypothalamus there are a number of other brain areas that have important roles in controlling the release of hormones from the pituitary gland. The interaction between the autonomic nervous system and the limbic system will be discussed in a later section but at this point note that the complex limbic systems are thought to have extensive influence upon the release of pituitary hormones. These effects are exerted either directly or indirectly via influence upon the hypothalamus.

One point of note before the end of this section is that Folkow and von Euler (1954) reported that electrical stimulation of different areas of the posterior hypothalamus of cats resulted in the differential release of adrenaline and noradrenaline from the adrenal medulla. This important finding does not appear to have been followed up or replicated by other workers.

From this short section on the role of the endocrine system it can be seen how closely the autonomic and endocrine systems interact. To consider the two systems as operating independently from each other is to present a grossly over-simplified account of the physiological system; one that is a far cry from the dynamic interplay that occurs inside the nervous body.

BODY TEMPERATURE AND THE HYPOTHALAMUS

Analysis of how the body temperature is maintained by the brain presents a good example of an interaction between the hypothalamus and the autonomic nervous system, an interaction in which the sympathetic system plays a considerable role. Results obtained mainly from animal studies show that the anterior regions of the hypothalamus are concerned with adjustments to increases in bodily heat while posterior regions are concerned with adjustments for decreases in bodily temperature. Although there has been a considerable amount of fine-grained research since 1968, the overall conceptions concerning temperature regulation have not substantially altered since then (Cabanac, 1975), although Cabanac has put forward a case for a revision of certain conceptions.

Normal body temperature is regulated at 37°C and, as Feldberg (1970) has noted, it remains remarkably constant over the whole life-span. Physiologists have shown that whereas the outer core temperature of the peripheral bodily tissues varies considerably, inner core temperature of the vital internal organs is maintained within very narrow limits. Warming or cooling the body results in peripheral thermoreceptors signalling to thermosensitive areas in the hypothalamus. These neural areas serve to integrate both central and peripheral effects. Temperature-sensitive areas of the hypothalamus can either be stimulated by peripheral thermoreceptors of chances in the temperature of the blood circulating in the brain.

Using a single cell recording technique Hellon (1967) showed that 23 of 227 nerve cells in the anterior hypothalamus showed temperature sensitivity. Of the 23 thermosensitive neurones, 17 showed increases in firing rate related to increases in temperature and 6 showed increases in firing rate related to decreases in temperature. The temperature-sensitive cells operated over a range of 38.5°C to 39.5°C.

The body appears to have a number of parallel negative feedback loops effecting a hypothalamic 'set point'. Alteration of the 'set point' in either direction brings opposing thermoregulatory mechanisms into play. The 'set point' should not be thought of as an actual temperature scale value but a point at which hypothalamic temperature control processes come into play. The 'set point' may undergo changes in threshold values as in the case of fever where it may be raised to an abnormally high threshold.

Inevitably, in any system as complex as temperature regulation, heat production by the body and the management of heat loss comprise interlocking systems. In order to keep this account relatively simple we will consider heat loss and heat gain mechanisms separately. Heat loss from the body is by radiation, evaporation, conduction, and convection, or a combination of these processes. Heat production is by: (1) increases in the basal metabolism; (2) shivering; (3) hormonal effects; (4) sympathetic effects.

In the normal way body heat loss is controlled by constriction of the blood vessels in the skin creating an effective insulator to maintain the vital 'inner core' temperature. If heat loss from the body becomes too great to be controlled by vasoconstriction, a large number of animals are able to gain additional insulation by fluffing out their fur or feathers (piloerection). The two processes just mentioned are under the control of the sympathetic nervous system. The next stage in hypothermia of the body is the brain controlled mechanism of shivering. The rapid and rhythmic contraction of the somatic muscles seen in the shivering person produces an approximately fivefold increase in body heat from increases in metabolic activity. The problem of temperature control in a body losing too much heat is not just dealt with by increases in cutaneous insulation or shivering because additional heat is produced by the endocrine system. For example, in low temperature conditions adrenaline released from the adrenal medulla produces a general increase in cellular metabolism and mobilizes glycogen stored in the liver. ACTH causes the release of the corticoids from the adrenal cortex and the thyrotrophic hormone causes the thyroid gland to release thyroxine. Both of these important pituitary processes play critical roles in bodily temperature maintenance.

In addition to its role in heat loss over the normal range the autonomic nervous system has recently been shown to play an important role in two special instances of body heat loss. The first is called nonshivering thermogenesis (Jansky, 1971). Nonshivering thermogenesis occurs through a chronic increase in the output of adrenal catecholamines that occurs in warm blooded animals exposed to low temperatures for prolonged periods. It has been estimated that the increased metabolic processes induced by the elevated levels of catecholamines produces a twofold increase in body heat. The second special instance relates to the general problem presented by the fact that a new born mammal has a large surface area from which heat can be lost. Until normal growth and development decreases the ratio between the surface and body volume, neonates, especially if small prematures, experience a vulnerable period for body temperature regulation. During this critical period neonatal deposits of 'brown fat' act as special energy repositories which supplement other methods of heat production in the body. Release of energy from 'brown fat' is controlled by the catecholamine noradrenaline (Smith and Horovitz, 1969).

Heat is continuously produced in the body as a by-product of metabolism. Although metabolic processes have been mentioned at a number of points in this section it is not proposed to deal with them in any detail. A good introduction to metabolism and its relationship to body heat will be found in Guyton (1971). In the normal way loss of heat from the body is controlled by dilation of the blood vessels in the skin; the additional volume of blood is cooled by being circulated in the peripheral skin. If an animal is unable to lose sufficient heat by this mechanism of vasodilation, panting or sweating occurs. Panting is the mechanism used by the majority of animals to reduce unwanted body temperature, but as pointed out in Chapter 1 panting is clearly unadaptive in humans where oral communication was to become so important. Evaporation of sweat from the surfaces of the body enables considerable latent heat to be carried away. Under tropical conditions humans may excrete several litres of sweat during the course of a day.

Hypothalamic neurochemicals play important roles in the thermoregulatory processes and Feldberg has suggested that cells in the anterior region are influenced by adrenergic pathways releasing noradrenaline and serotonin. Feldberg's view is that two types of cells exist, having opposing actions; for example, serotonin induces shivering while noradrenaline abolishes shivering. However, as the effects appear to be species-specific the exact mechanism is far from clear. Injections of serotonin into the hypothalamus of rabbits produces a biphasic response. An initial reaction of shivering results in body temperature increasing. However, administration of serotonin into the hypothalamus of a rat or mouse produces a straight fall in body temperature without the subsequent increase found in rabbits. This species difference may be due to an interaction between the body / surface ratio of different species and the amount of chemical injected. Biogenic amines have also been shown to be affected by the ambient temperature surrounding the experimental animal. Additional complications arise from the fact that acetylcholine has a marked effect and also produces hypothermia and it has been suggested that it plays an active role in producing hypothermia (Feldberg, 1970). Clearly, from what has been said previously in this book about the role of neurochemicals and neurohormones in the processes of the hypothalamus, it would be surprising if they did not have significant roles in temperature regulation but we seem to be some way from understanding the exact details.

Before we conclude this section, there is a further aspect that we should consider. In addition to our physiological mechanisms for temperature control, we may use behavioural means to control our body temperature. This may involve removing certain clothing when we feel hot or putting more on when we feel cold, to turning up the central heating temperature thermostat of the room or taking a cold shower. Weiss and Laties (1961) have shown that animals can learn to regulate their temperatures by behavioural means. As indicated in Chapter 1, certain reptiles need to spend considerable proportions of their lives behaviourally manipulating their body temperatures. In general terms, humans spend a relatively small proportion of their time in behavioural manipulation of body temperature.

Cognitive factors have been shown to apparently affect body temperature; for example, consider the situation when we see a poster advertising a holiday resort by means of a sunny beach scene. In this situation we sense something of the warmth of the scene. How does this come about? Is it the result of a conditioned response of vasodilation that we would experience in the actual situation or merely a feeling? Conversely, does a cold scene produce vasoconstriction of the peripheral blood vessels? Very little research appears to have been carried out on this type of problem. Emotional factors also produce changes in body temperature but these will be considered later. The importance of these points arises from the fact that if conditioned responses of the autonomic nervous system can be shown to operate in such situations we might gain valuable insights into the mechanisms of higher brain centres.

THE THALAMUS

The thalamus is the 'sensorium' of the brain and, like the limbic system that will be discussed in the next section, it is complex and its relationships with the autonomic nervous system are important but not always obvious. The thalamus forms the largest part of the diencephalon, which consists of the thalamus, epithalamus, hypothalamus, and subthalamus. The two halves of the thalamus, situated within the two cerebral hemispheres vary in size and there is marked intra-individual variation as well as considerable species differences. The size differences of the thalamus in humans are mainly due to variations in the dorsally sited pulvinar region that has been shown in a number of clinical studies (Watts, 1975) to play an important role in the storage and retrieval of long-term memory. From its central position, see Figure 3.9, the thalamus receives virtually all of the sensory information being passed into the brain. This sensory information is then passed on to the cerebral cortex via thalamic nuclei projections. Thalamic projections are found in all the main cortical lobes, see Figure 3.9, and thalamic nuclei receive cortical innervation in return.

In general, sensory information enters the brain via two main routes. Peripheral information is carried in the spinal cord via the somatosensory pathway and we may suppose that a considerable volume of sensory information from the autonomic system travels in these pathways. The second group arrives at the thalamus via the sensory cranial nerves. The thalamus is commonly divided into three nuclear masses according to cortical interconnections. The sensory relay nuclei of the thalamus consist of the lower part of the ventrobasal complex, a posterior part of this complex, together with the lateral geniculate nucleus and the medial geniculate nucleus. The medial geniculate nuclei bodies are part of the auditory pathways and the lateral geniculate bodies receive visual information direct from the retinas. The anterior areas of the ventrobasal complex receive projections from the

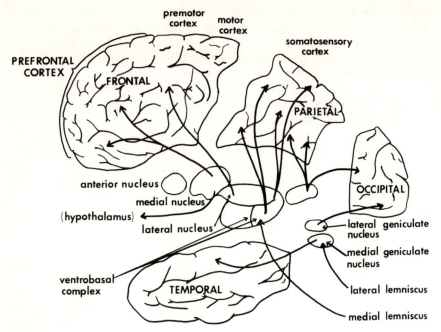

premotor cortex
motor cortex
somatosensory cortex
PREFRONTAL CORTEX
FRONTAL
PARIETAL
anterior nucleus
medial nucleus
(hypothalamus)
lateral nucleus
OCCIPITAL
lateral geniculate nucleus
medial geniculate nucleus
ventrobasal complex
TEMPORAL
lateral lemniscus
medial lemniscus

Figure 3.9   Connections between the thalamus and cortical areas of the brain. From its central position the thalamus is well suited to receive and pass projections to each of the four main cortical areas. It is thought to monitor all sensory information entering the brain. From *An Introduction to Physiological Psychology*, by Allen M. Schneider and Barry Tarshis. Copyright © 1975 by Random House, Inc. Reprinted by permission of Random House, Inc.

cerebellum and the extrapyramidal sensory motor tracts of the basal ganglia, these in turn project to the motor area of the cerebral cortex.

As mentioned previously, the association nuclei form the bulk of the thalamic masses and have extensive connections with the association cortex of the frontal lobes. The medial thalamic nuclei receive projections from other parts of the thalamus as well as the hypothalamus. The anterior nucleus is part of the limbic system that will be discussed next, it also receives projections from the mamillary bodies of the hypothalamus, see Figure 3.10. In addition to the sensory inputs mentioned in this section the reticular formation has extensive projections into the thalamus.

Clinical investigations have revealed many of the functions of the thalamus. Pathological disorders may affect consciousness or sleep and a wide range of moods (Watts, 1975). Such disturbances may result in patients suffering from a disproportionate amount of pain if their skin is pricked with a pin. Conversely, intractable pain experienced by patients suffering from terminal cancer illnesses has been relieved by allowing them to deliver minute electric shocks to themselves via electrodes that surgeons have positioned in their thalamic areas.

Watts (1975) states that normally the thalamus has primary control over most of the autonomic functions but in abnormal states, ranging from pathological to intense emotional states found in emergency situations, the hypothalamus takes over the primary control of autonomic functions.

THE LIMBIC SYSTEM

The limbic system was named by MacLean (1949) from the earlier concept by the French anatomist Broca, who in 1878 called *le grand lobe limbiqué*. It consists of a circle of complex brain structures lying immediately below the cerebral cortex. Like the thalamus it has connections with the separate lobes of the cerebral cortex and its influence is extensive. The limbic / hypothalamic interaction is sufficiently complex for Isaacson (1974) to include the hypothalamus as part of the limbic system. From the point of view being advanced in this book, attention is drawn to the fact that the limbic system has extensive connections with the hypothalamus and the endocrine system.

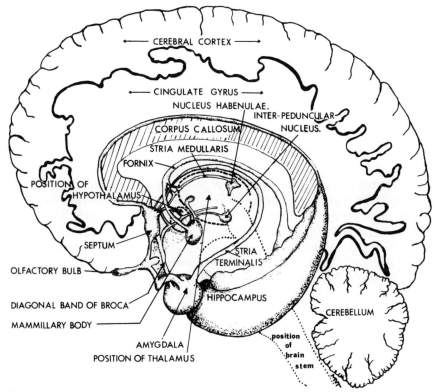

Figure 3.10 A schematic drawing of the limbic system. The limbic system consists of about 53 regions and 35 associated tracts. The main structures of the limbic system lie immediately below the cortex of the brain. One of its essential roles relates to the integration of the autonomic nervous system and the endocrine system.

The limbic system consists of 53 regions and 35 major nerve tracts (Watts, 1975). A schematic diagram showing the main parts and their relationship to the hypothalamus, thalamus and the cerebral cortex is shown in Figure 3.10. For more specialized information about its structure and function the interested reader will find the following references of value: Hockman (1972) contains a collection of chapters considering the interaction between the limbic and the autonomic nervous system; Isaacson (1974) has reviewed behavioural studies that have been carried out in attempts to elucidate the function of the limbic system; DiCara (1974) contains a selection of research topics; Watts (1975) presents a personal comprehensive and integrated account related to clinical studies.

Broca noted that the limbic system had a relationship with the olfactory lobes and that a limbic system was found in the brains of all mammals. The nervous pathway projections from the olfactory lobes led to the limbic system being considered as exclusively concerned with the processes of smell and in older textbooks it is often called the rhinencephalon or 'smell' brain. It was not until Papez (1937), in a theoretical paper, pointed out its role in mood and emotion that researchers began to look at other possibilities. Clearly, the limbic system has a role in olfaction but it could not be considered to be one of its primary roles.

Although each part of the limbic system appears to have a primary function the interrelationships between the different parts makes interpretation very difficult. The 53 regions and the supporting 35 tracts are constantly switching and changing in terms of nervous activity; any area may be found to be active over a wide range of behaviour that may appear to be quite unrelated. This suggests that a significant dimension of its activity is qualitative rather than quantitative. The complexity of the limbic system may explain many human individual variations in terms of amount of sleep needed or sensitivity to pain. The complexity of the limbic system probably renders it particularly vulnerable to pathological conditions and makes it sensitive to a wide range of drugs.

The same complexity that makes interpretation of research findings so difficult has stimulated theoretical papers from a number of research workers. It is of interest to attempt to summarize these and look for common themes that will enable us to begin to suggest how the limbic and the autonomic systems work together. MacLean (1975) has put forward a theoretical conception of the brain in which he conceives the brain as having three distinct layers. The layers of his triune brain are: (1) the protoreptilian brain which consists of the hindbrain and brain stem; (2) the palaeomammalian brain which consists primarily of limbic system structures; (3) the neomammalian brain which consists of the neocortex. The protoreptilian brain is concerned with survival mechanisms. The palaeomammalian brain is primarily concerned with visceral mechanisms but it contains a rudimentary 'self-awareness'. The neomammalian is the part of the brain that is concerned with the processes of cognition and making fine-grain analysis of the external world; it is the part of the brain that is able to attempt the prediction of future events. The neomammalian brain is largely unaffected by the internal world controlled by the other two layers of the brain. It is the palaeomammalian part of the brain or the limbic system that acts to integrate the neomammalian and the protoreptilian layers of the nervous system.

Vanderwolf (1971) suggested a conception of the brain having three parts with the coupling between 'conscious' and 'unconscious' movement being made by the limbic system. Vanderwolf suggests that there is a one brain mechanism for voluntary movements and one for automatic movements found in postural and other movements that are usually regarded as 'unconscious'. For example, the initial feelings of cold in animals are not usually associated with voluntary movement (behavioural thermoregulation) but with cutaneous vasoconstriction, shivering and a huddled posture. The point being made is that the 'conscious' or high level of awareness mechanisms are largely mediated by the highest cortical functions of the brain whereas the 'unconscious' or low level of awareness mechanisms are mediated by the mid and hindbrain structures with a minimum of cortical control. The function of the limbic system is to link the 'drive' or motivational regions of the mid and hindbrain to the voluntary movement centres located in the cerebral cortex. Normally the two parts are quite synchronous but in certain pathological states or situations of uncertainty the two parts are not in synchrony and dissociate. This dissociation can produce delay in response until the uncertainty is resolved either by a situational change or a choice of the alternatives. Resolution of uncertainty allows for the normal smooth integration between the motor systems of the autonomic and the somatic nervous systems to proceed in its normal fashion.

How integration occurs between the motor systems of the autonomic and the somatic appears to be a neglected area of study that may yield valuable insights into the relationships between autonomic and limbic system structures.

Isaacson (1974), arguing from evidence obtained from behavioural studies, feels that the limbic system essentially has an inhibiting role. Analysis of the learning processes of animals suggests that an essential behavioural difference between the mammals and lower orders is that mammals are able to forget more effectively. Isaacson points out that the basic mechanisms for learning and memory are contained in brain structures below the neocortex. Also, stimulation of limbic system regions often produces suppression of ongoing behaviour; conversely lesions made in the limbic system often seem to release activity. From this type of argument Isaacson concludes that the essential role of the limbic system is suppression of the activities in what MacLean called the protoreptilian brain.

Watts (1975) states that the primary role of the limbic system is to interpret the total sensory input as pleasant or unpleasant. It makes instant decisions after evaluating brain input at any time as either innocuous or dangerous; it then directs the integrated somatic activity into forward or backward movement. These approach-avoidance somatic movements require correlated activity by the autonomic nervous system and the endocrine systems.

It seems that an important role for the limbic system is to integrate autonomic and endocrine responses. As mentioned at a number of earlier points the autonomic nervous system does not suddenly switch in during emergency states but is continuously being adjusted to allow for body movement and temperature changes etc. The limbic system appears to direct the normally harmonious total physiological reaction and it does this by orchestrating the motor systems of the autonomic and the somatic nervous systems and backs up these systems with appropriate endocrine responses.

THE CEREBELLUM

In view of the above emphasis upon somatic-autonomic interaction, it should come as no surprise to discover that a number of workers have reported that stimulation of the cerebellum can produce autonomic effects (Moruzzi, 1940; Dow, 1961). Moruzzi showed that stimulation of the anterior lobe of the cerebellum produced an inhibitory effect on vasopressor and cardiovascular reflexes as well as having an effect upon respiration. Bard *et al.* (1947) claimed that the nodulus area of the cerebellum was essential to the production of motion sickness in dogs. Wang and Brown (1956) have shown inhibition of the galvanic skin response by stimulating the anterior lobe of the cerebellum.

In a study that was primarily designed to show sensorimotor responses from stimulation of the cortical motor tracts (pyramidal tracts) Landau (1953) demonstrated autonomic responses. He suggested that some of the pyramidal tracts subserve autonomic functions and, in line with the thesis put forward in this book, argued that the predominant effect of the pyramidal tract upon the autonomic nervous system was one of facilitation.

THE FRONTAL LOBES

Periodically over the years, researchers have drawn attention to the fact that some control centres for autonomic function are located in the frontal lobes of the brain. Fulton (1951) stated that the frontal lobes represent the highest level of autonomic control. One of Fulton's students (Kennard, 1949) reported alterations in gastrointestinal functions, temperature regulation, and blood flow following removal of the frontal lobes, Wall and Davis (1951) have confirmed this. Brutkowski (1964, 1965) has pointed to changes in fear and sexual behaviour following removal of the frontal cortex in dogs. Amassian (1951) has demonstrated autonomic control from cortical centres and Schlag and Scheibel (1967) have indicated the presence of inhibitory mechanisms in the forebrain that affect autonomic activity. Le Gros Clark (1948) directly challenged the long standing contention, that the autonomic nervous system was not represented in the frontal lobes, by arguing that the frontal lobes represented an integration centre for the cognitive processes receiving impulses from the viscera, either directly via the hypothalamus or indirectly via the thalamus. However, these findings appear to have been consistently overlooked. This point is well illustrated in a recent review (Nauta, 1971) which consists largely of a detailed review of cognitive functions, with little consideration of autonomic bodily systems.

For many years the frontal lobes were spoken of as the silent areas, as electrical stimulation to these areas frequently failed to produce responses. Stanley and Jaynes (1949) have suggested that an inverse relationship exists between the rate of stimulation and the response from the frontal lobes and this fact has masked the frontal lobe / autonomic relationship. For example, Delgado and Livingston (1948) found effects of electrical stimulation on a respiratory centre in the frontal lobes showing this inverse function. A stimulation rate of 180 Hz produced no effect while 6 Hz produced a dramatic respiratory arrest. In part, it is this type of inverse relationship which may have masked autonomic functions of the frontal lobes and it is reminiscent of the point recalled earlier, by Pick (1970), that characteristics of the autonomic nervous system may show qualitative as well as quantitative differences when compared to the somatic nervous system.

The function of the frontal lobes is integration of information, not only from cortical lobes and external world but also from the internal world of the autonomic nervous system. The frontal lobes initiate the total bodily pattern for any required response; the limbic system integrates necessary somatic and visceral changes. Any alterations in the somatic system require autonomic nervous system changes to back them up; if I move away from my desk, changes in blood distribution are required to take account of my muscle movements. Consideration of this statement suggests that we can expect patients with pathological damage to their frontal lobes to show malfunction of normally smooth integration of somatic and autonomic functions. It is this dissociation of somatic and autonomic brain processes that is responsible for the difficulties these patients have in carrying out tasks involving delayed responses. These patients reveal small but significant timing delays, signifying that the normally smoothly integrated response pattern has been disturbed. Their delayed responses are produced by the 'reorganization' processes these patients need to make. This point is supported by Holloway and Parsons (1972) who, using a simple reaction time task that involved a variable time signal, found that brain-damaged patients, unlike control subjects, displayed either no relationship or an inverse relationship between autonomic and somatic activity. In a later paper, Lovallo *et al.* (1973) showed reduced levels of arousal in brain damaged patients compared to control patients.

During the last three decades there have been considerable changes in our conceptions about the relationship of the brain to the autonomic nervous system. In the 1940s the hypothalamus was considered to control the autonomic nervous system and was held to be responsible for most of the interesting motivational behaviours. This overstatement of its role appears to have arisen because of a confusion over what we might call the 'executive' and 'policy' areas of the brain. The role of the hypothalamus, if anything, is 'executive' and it carries out the 'policy' of higher brain structures. If we make this assumption, we can suppose that because the 'policy' decisions of the brain in most cases are not stated in great detail, this allows for considerable latitude in the interpretation of decisions made by the higher brain centres. At the present time, as the complex roles of the thalamus and the limbic systems become to be understood more clearly, we can doubt such a purely 'executive' function for the hypothalamus. Recently we have come to think of the function of the brain in holistic terms. Pribram (1969) has advanced a view of the brain that is based on an analogy with photographic holograms where total information is stored in patterns that are faithfully reproduced even in a small fragment of the original plate. Willshaw *et al.* (1969) conceive brain function in terms of an interconnecting network of nuclei. This type of conception enables each cluster of nuclei to have a specific function but also has the advantage of preserving the undoubted specificity of many brain areas. It would give maximum protection against damage as any node would be able to be preserved in part by the surrounding network despite the node itself being destroyed.

Recent conceptions concerning the function of the brain have dealt in generalizations and few attempts have been made to relate the different functions of the brain to each other. As we have seen, the autonomic nervous system interrelates to most areas of the brain and this is what we should expect. The evolving brain did not 'abandon its older' functions. At each new stage important elements of previous functions were carried forward and integrated into the new structures. In terms of brain function we see the frontal cortex dealing with cascades of information from all parts of the brain: cascades from other cortical lobes, a thalamic cascade, a limbic system cascade, and an autonomic nervous system cascade, and a endocrine cascade. Within the cascade of each major system we may suppose that there are smaller cascades, all of which are capable of modifying brain function. Somehow this information is assessed and modified in an apparently democratic way. The question, How? remains a scientific challenge for the future.

REFERENCES

Amassian, V. E. (1951). 'Cortical representation of visceral afferents', *J. Neurophysiol.,* **14**, 445-50.

Anlezark, G.M., Crow, T.J. and Greenway, A.P. (1973). 'Impaired learning and decreased cortical norepinephrine after bilateral locus coerulus lesions', *Science,* **181**, 682-4.

Arians-Kappers, C.V., Huber, G.C. and Crosby, E.C. (1960) *The Comparative Anatomy of the Nervous System of Vertebrates Including Man* (3 volumes). Hafner, New York.

Bard, P., Woolsey, C. N., Snider, R. S., Mountcastle, V. B. and Bromiley, R. B. (1947). 'Delimitation of central nervous mechanisms involved in motion sickness', *Fed. Proc.,* **6**, 72.

Baust, W. and Niemczyk, H. (1963). Studies on the adrenaline-sensitive component of the mesencephalic reticular formation. *J. Neurophysiol.,* **26**, 692–704.

Beagley, W. K. and Holley, T. L. (1977). 'Hypothalamic stimulation facilitates contralateral visual control of a learned response', *Science,* **196**, 321-2.

Bernard, C. (1877). *Lecons sur le Diabete et la Glycogenese Animal.* Bailliere, Paris.

Bindra, D. (1969). 'A unified interpretation of emotion and motivation', *Ann. New York Academy Sciences,* **159**, 1071-83.

Blaschko, H. Muscholl, E. (Eds.) (1972). *Handbook of Experimental Pharmacology Volume 33. Catecholamines.* Springer Verlag, Berlin.

Bonvallet, M., Dell, R. and Hugelin, A. (1954). 'Influence de l'adrenaline sur le controle reticulaire des activites corticale et spinale', *J. Physiol. (Paris)*, **46**, 262-5.

Brutkowski, S. (1964). 'Prefrontal cortex and drive inhibition', in T.M. Warren and K. Akert (Eds.) *The Frontal Granular Cortex and Behaviour*. McGraw-Hill, New York.

Brutkowski, S. (1965). 'Functions of the prefrontal cortex in animals', *Physiol. Rev.*, **45**, 721-46.

Buck, R. (1976). *Human Motivation and Emotion*. John Wiley and Sons, London.

Cabanac, M. (1975). 'Temperature regulation', p. 415-439 in J.H. Comroe (Ed.) *Annual Review of Physiology Volume 37*, Annual Reviews Inc.

Cannon, W.B. (1934). 'Hunger and thirst', in C. Murchison (Ed.) *Handbook of General Experimental Psychology*. Clark University Press.

Cannon, W.B. and Washburn, A.L. (1912) 'An explanation of hunger', *Amer. J. Physiol.*, **29**, 441-54.

Clark, Le Gros, W.E. (1938). 'Morphological aspects of the hypothalamus', in W.E. Le Gros, Clark, J. Beattie, G. Riddock and N.M. Dott (Eds.) *The Hypothalamus: Morphological, Functional, Clinical, and Surgical Aspects*. Oliver and Boyd, Edinburgh (1938).

Clark, Le Gros, W.E. (1948). 'The connections of the frontal lobes of the brain', *Lancet*, **1**, 353-6.

Delgado, J.M.R. and Livingstone, R.B. (1948). 'Some respiratory, vascular, and thermal responses to stimulation of orbital surfaces of the frontal lobe', *J. Neurophysiol.*, **11**, 39-55.

Dell, P., Bonvallet, M. and Hugelin, A. (1954) 'Tonus sympathique adrenaline et controle reticulaire de la motricite spinale', *EEG Clin. Neurophysiol.*, **6**, 599-618.

DiCara, L.V. (1974). *Limbic and Autonomic Nervous Systems Research*. Plenum Press, New York.

Dow, R.S. (1961). 'Some aspects of cerebellar physiology', *J. Neurosurgery*, **18**, 512-30.

Feldberg, W. (1970). 'Monoamines of the hypothalamus as mediators of temperature response', in L. Martini, M. Motta and F. Fraschini (Eds.) *The Hypothalamus*. Academic Press, New York.

Folkow, B. and von Euler, U.S. (1954). 'Selective activity of noradrenaline and adrenaline producing cells in the cat's suprarenal gland by hypothalamic stimulation. *Circulation Res.*, **2**, 191-5.

Fulton, J.F. (1951). *Frontal Lobotomy and Affective Behaviour: A Neurophysiological Anlaysis*. Chapman Hall, London.

Fulton, J.F., Ranson, S.W. and Frantz, A.M. (1940). *The Hypothalamus and Central Levels of Autonomic Function*. Williams and Wilkins, Baltimore.

Gardner, E.D. (1975). *Fundamentals of Neurology* (6th edn.). W.B. Saunders, New York.

Gellhorn, E. (1943). *Autonomic Regulation: Its Significance for Physiology, Psychology, and Neuropsychiatry*. Interscience, New York.

Gellhorn, E. (1957). *Autonomic Imbalance and the Hypothalamus*. University of Minnesota Press.

Gellhorn, E. (1967). 'The tuning of the nervous system: physiological foundations and implications for behaviour', *Perspectives in Biology and Medicine*, **10**, 559-619.

Germana, J. (1969). 'Central efferent processes and autonomic-behavioural integration', *Psychophysiol.* **6**, 78-90.

Gloor, P. (1954). 'Autonomic functions of the diencephalon: a summary of the experimental work of Professor W. R. Hess', *Amer. Med. Assoc. Arch. Neurol. Psychiat.*, **71**, 773-90.

Grossman, S.P. (1973). *Essentials of Physiological Psychology*. Wiley, London.

Guyton, A.C. (1971). *Textbook of Medical Physiology*. W. B. Saunders, New York.

Hartman, B.K., Zide, D. and Udenfried, S. (1972). 'Hydroxylase as a marker for the central noradrenergic nervous system in rat brain', *Proc. National Academy of Sciences (USA)*, **69**, 2722-6.

Hellon, R.F. (1967). 'Thermal stimulation of hypothalamic neurones in unanaesthetized rabbits', *J. Physiol.*, **193**, 381-95.

Hess, W.R. (1954). *Diencephalon: Autonomic and Extrapyramidal Functions.* Heinemann, London.

Hockman, C.H. (ed.) (1972). *Limbic System Mechanisms and Autonomic Functions.* C.C. Thomas, Springfield.

Holloway, F.A. and Parsons, O.A. (1972). 'Physiological concomitants of reaction time performance in normal and brain-damaged subjects', *Psychophysiology,* **9**, 189-98.

Hornykiewicz, O. (1973). 'Dopamine in the basal ganglia', *Brit. Med. Bull.*, **29**, 172-8.

Horovtiz, Z.P., Beer, B., Clody, D.E., Vogel, J.R. and Chasin, M. (1972). 'Cyclic AMP and anxiety', *Psychosomatics*, **13**, 85-92.

Isaacson, R.L. (1974). *The Limbic System.* Plenum Press, New York.

Iversen, L.L. (1967). *The Uptake and Storage of Noradrenaline in Sympathetic Nerves.* Cambridge University Press.

Iversen, L.L. (ed.) (1973). Catecholamines. *Brit. Med. Bull.*, **29**.

Iverson, L.L., Iverson, S.D. and Snyder, S.H. (1975). *Handbook of Psychopharmacology Volumes 1-6.* Plenum Press, New York.

Jansky, L. (1971). *Nonshivering Thermogenesis.* Swets and Zeitlinger, Amsterdam.

Jouvet, M. (1969). 'Biogenic amines and the states of sleep', *Science*, **163**, 32-40.

Kappers, J.A. (1965) 'Survey of innervation of epiphysis cerebri and accessory pineal organs of vertebrates', in J.A. Kappers and J.P. Schade (Eds.) *Progress in Brain Research* Volume 10. Elsevier, Amsterdam.

Kappers, A.J. and Schade, J.P. (eds.) (1965). 'Structure and function of the epiphysis cerebri', in *Progress in Brain Research* Volume 10. Elsevier, Amsterdam.

Kappers, J.A., Smith, A.R. and De Vries, R.A.C. (1974). 'The mammalian pineal gland and its control of hypothalamic activity', in D.F. Swaab and J.P. Schade (Eds.) *Integrative Hypothalamic Activity.* Elsevier, Amsterdam.

Karplus, J.P. and Kreidl, A. (1909). 'Gehirn und Sympathicus: 1 Zwischenhirnbasis and Halsympathicus', *Pfluger's Archives ges Physiologie*, **129**, 138-44.

Karplus, J.P. and Kreidl, A. (1910). 'Gehirn and Sympathicus: 2 ein Sympathicuszentrum in Zwischenhirn', *Pfluger's Archives ges Physiology*, **135**, 401-16.

Karplus, J.P. and Kreidl, A. (1912). 'Gehrin and Sympathicus: 3 Sympathicausleitung im Gehirn and Halsmark', *Pfluger's Archives ges Physiologie*, **143**, 109-27.

Kempf, E.J. (1920). *Psychopathology*, Mosby, St. Louis.

Kennard, M. (1949). Autonomic function in the pre-central motor cortex, in P.C. Bucy (Ed.) *The Pre-central Motor Cortex.* University of Illinois Press.

Landau, W.M. (1953). 'Autonomic responses mediated via corticospinal tract', *J. Neurophysiol.*, **16**, 299-311.

Leshner, A.I. (1978). *An Introduction to Behavioural Endocrinology.* Oxford University Press, London.

Livett, B.G. (1973). 'Histochemical visualization of adrenergic neurones', in L.L. Iversen (Ed.) *Catecholamines. Brit. Med. Bull.*, **29**, 93-9.

Lovallo, W., Parsons, O.A. and Holloway, F. (1973). 'Autonomic arousal in normal, alcoholic, and brain damaged subjects as measured by the plethysmograph response to cold pressor stimulation', *Psychophysiol.*, **10**, 166-76.

MacDonnell, M.F. and Flynn, J.P. (1966). 'Control of sensory fields by stimulation of the hypothalamus', *Science*, **152**, 1406-8.

MacLean, P.D. (1949). 'Psychosomatic disease and the "visceral brain": recent developments bearing upon the Papez theory of emotion', *Psychosomatic Medicine*, **11**, 338-53.

MacLean, P.D. (1975). 'Sensory and perceptive factors in emotional functions of the triune brain', in L. Levi (Ed.) *Emotions: Their Parameters and Measurements*. Raven Press, New York.

Marshall, J.P. (1975). 'Increased orientation to sensory stimuli following medial hypothalamic damage in the rat', *Brain Res.*, **86**, 373-87.

Marshall, J.P., Turner, B.H. and Teitelbaum, P. (1971). 'Sensory neglect produced by lateral hypothalamic damage', *Science*, **174**, 523-5.

Marshall, J.P. and Teitelbaum, P. (1974). 'Further analysis of sensory inattention following lateral hypothalamic damage in rats', *J. comp. physiol. Psychol.*, **86**, 375-95.

Martini, L., Motta, M. and Fraschini, F. (Eds.) (1970). *The Hypothalamus*. Academic Press, New York.

Mess, B., Zanisi, M. and Tima, L. (1970) 'Site of production of releasing and inhibiting factors', in L. Martini, M. Motta and F. Fraschini (Eds.) *The Hypothalamus*. Academic Press, New York.

Moore, R.Y., Heller, A., Bhatnager, R.K., Wurtman, R.J. and Axelrod, J. (1968). 'Central control of pineal gland: visual pathway', *Arch. Neurology*, **18**, 208-18.

Morgan, C.T. (1965). *Physiological Psychology*. McGraw-Hill, New York.

Moruzzi, G. (1940). 'Paleocerebellar inhibition of vasomotor and respiratory corotid sinus reflexes', *J. Neurophysiol.* 3, 20-32.

Motta, M. Crosignani, P.G. and Martini, L. (1975). *Hypothalamic Hormones*, Academic Press, New York.

Nagatsu, T. (1973). *Biochemistry of Catecholamines*. University Park Press, Baltimore.

Nauta, W.J. (1971). 'The problem of the frontal lobe: a reinterpretation', *J. Psychiatric Res.*, **8**, 167-87.

Obal, F. (1966). 'The fundamentals of the central nervous system control of vegetative homeostasis', *Acta Physiol. Academy Science* (Hungary), **30**, 15-29.

Papez, J.W. (1937). 'A proposed mechanism of emotion', *Arch. Neurol. Psychiat.*, **38**, 725-43.

Pick, J. (1970). *The Autonomic Nervous System: Morphological, Comparative, Clinical, and Surgical Aspects*. Lippincott, New York.

Pribram, K.H. (Ed.) (1969). *On the Biology of Learning*. Harcourt, Brace, and World, London.

Ranson, S.W. and Clark, S.L. (1953). *The Anatomy of the Nervous System* (9th edition). Saunders, Philadelphia.

Reynolds, R.W. (1965). 'An irritative hypothesis concerning the hypothalamic regulation of food intake', *Psychol. Rev.*, **72**, 105-16.

Romijn, H.J. (1973). 'Parasympathetic innervation of the rabbit pineal gland', *Brain Res.*, **55**, 431-6.

Rothballer, A.B. (1956). 'Studies on the adrenaline-sensitive component of the reticular activating system', *EEG. Clin. Neurophysiol.*, **8**, 603-21.

Schapiro, S. and Salas, M. (1971). 'Effects of age, light, and sympathetic innervation on electrical activity of the rat pineal gland', *Brain Res.*, **28**, 47-55.

Scharrer, E. and Scharrer, B. (1963). *Neuroendocrinology*. University of Columbia Press.

Schlag, J. and Scheibel, A. (Eds.) (1967). Forebrain inhibitory mechanisms. *Brain Research*, **6**, 1-198.

Schneiderman, N., Francis, J., Sampson, L.D. and Schwaber, J.S. (1974). 'Central nervous system integration of learned cardiovascular behaviour', in L.V. DiCara (Ed.) *Limbic and Autonomic Nervous Systems Research*. Plenum Press, New York.

Smelik, P.G. (1970). 'Integrated hypothalamic response to stress', in L. Martini, M. Motta and F. Fraschini, (Eds). *The Hypothalamus* Academic Press, London.

Smith, R.E. and Hortwitz, B.A. (1969). 'Brown fat and thermogenesis', *Physiol. Rev.*, **49**, 330-425.

Stanley, W.C. and Jaynes, J. (1949). 'The function of the frontal cortex', *Psychol. Rev.*, **156**, 18-32.

Stellar, E. (1954). 'The physiology of motivation', *Psychol. Rev.*, **61**, 5-22.

Teitelbaum, P. and Epstein, A.N. (1962). 'The lateral hypothalamic syndrome: recovery of feeding and eating after lateral hypothalamic lesions', *Psychol. Rev.*, **69**, 74-90.

Udenfriend, S. (1962). *Fluorescence Assay in Biology and Medicine and Molecular Biology* Academic Press, New York.

Ungerstedt, U. (1971). 'Stereotaxic mapping of the monoamine pathways in the rat brain', *Acta Physiologica Scandinavica* Suppl **367**.

Usdin, E. and Snyder, S. (Eds.) (1973). *Frontiers in Catecholamine Research. Proc. of the Third International Catecholamine Symposium.* Pergamon, London.

Vanderwolf, C.H. (1971). 'Limbic-diencephalic mechanism of voluntary movement. *Psychol. Rev.*, **78**, 83-113.

Vogt, M. (1959). 'Catecholamines in the brain', *Pharmacol. Rev.*, **11**, 483-9.

Vogt, M. (1973). 'Functional aspects of catecholamines in central nervous systems', *Brit. Med. Bull.*, **29**, 168-71.

Wall, P.D. and Davis, G.D. (1951). 'Three cerebral cortical systems affecting autonomic function', *J. Neurophysiol.*, **14**, 507-17.

Wang, G.H. and Brown, V.N. (1956). 'Suprasegmental inhibition of an autonomic reflex', *J. Neurophysiol.*, **19**, 564-72.

Warburton, D.M. (1975). *Brain Behaviour and Drugs: Introduction to the Neurochemistry of Behaviour.* J. Wiley and Son, London.

Watts, G.O. (1975). *Dynamic Neuroscience: Its Application to Brain Disorders.* Harper and Row, New York.

Weiss, B. and Laties, V.G. (1961). 'Behavioural thermoregulation', *Science*, **133**, 1338-44.

Willshaw, D.J., Buneman, O.P. and Longuet-Higgins, H.C. (1969). 'Non-holographic associative memory', *Nature*, **222**, 960-2.

Wooley, D.W. (1962). *The Biochemical Basis of Psychoses.* Wiley, New York.

Wooley, D.W. (1967). 'Involvement of the hormone serotonin in emotion and mind', in D.C. Glass (Ed.) *Neurophysiology and Emotion.* Rockefeller University Press.

Wurtman, R.J., Axelrod. J. and Kelly, D.E. (1968). *The Pineal* Academic Press, New York.

CHAPTER 4

Psychophysiology and Psychopharmacology

Psychophysiology is a relatively new science about an old topic. Although the influence of the autonomic nervous system on behaviour has been known and commented upon since antiquity, it was not until the development of the polygraph that the underlying physiological activity of the body could be permanently recorded and analysed in a systematic way. The Society for Psychophysiological Research was formed in 1962 and its official journal, *Psychophysiology*, first published in 1964. Sternbach (1966) published the first introductory textbook and Greenfield and Sternback (1972) the first handbook of psychophysiology. Details of specific techniques will be found in two manuals that were published in 1967 (Brown, 1967; Venables and Martin, 1967). An introduction to the psychophysiology of mental illness has been written by Lader (1975) which contains a useful introductory chapter concerned with the basic principles behind the most widely used techniques. Venables and Christie (1975) have edited a book in which contributors summarized recent research findings in a number of important areas of psychophysiology.

In general, a psychophysiologist tends to use bodily functions as his dependent variable and, as Sternbach emphasizes, workers in this field of psychology have always shown a firm interest in conditioning and factors that influence the autonomic nervous system. Autonomic measures commonly used by psychophysiologists are: sweating, various aspects of blood pressure and flow-rates, body temperatures, various measures of cardiac function, electrical characteristics of the skin, pupillary movement, and, in certain cases, gastric contractions.

Initially, psychophysiology showed a great deal of overlap with psychosomatic disease, see Roessler and Greenfield (1962), but during the last decade it has acquired a distinctive orientation that justifies its separate treatment from psychosomatic complaints. An early review of autonomic balance and function can be found in Darrow (1943); this article is republished in Sternbach (1966).

It must not be forgotten that psychophysiologists are also interested in measuring the electrical activity from most areas of the brain, muscular activity, and transducing a wide variety of physiological systems. However, these areas, apart from certain specific instances, lie outside the scope of this book and interested readers are referred to the references and textbooks cited, and the *Journal of Psychophysiology*. Clearly the field of psychophysiology is at a very crucial stage of development and there are many interesting problems that remain to be solved. For example, one of the most important in the near future must be how bodily rhythms are influenced by external factors. Investigation of this area will certainly involve the pineal-autonomic interaction mentioned in Chapter 3. The absence of a chapter on this area of psychophysiology is a notable omission from the handbook edited by Greenfield and Sternbach (1972).

Before we consider the theoretical problems of psychophysiology it will be useful to review briefly some of the most important tests of autonomic function that are used to determine the relative excitability of the sympathetic and the parasympathetic components. There are two main tests which have become standard procedures for determining the basic operating level of an individuals autonomic nervous system. Both involve measuring blood pressure changes, one uses drugs, the other uses cold water.

For a more exhaustive review the interested reader is referred to Monnier (1968) who gives a list of over 25 different tests of autonomic function.

THE MECHOLYL TEST

This test relates to an observation made by Altman *et al.* (1943) that the patterning of blood pressure following an injection of a parasympathetic drug called acetyl-beta-methylcholine (Mecholyl) related to certain psychiatric states. Normal patients when injected with the drug showed an initial fall in blood pressure which, under the influence of the sympathetic nervous system, would rise back to the base-line level. Autonomic dysfunction was indicated by an abnormal restoration of the blood pressure to the baseline level. This took the form of either a very rapid recovery with a marked overshoot indicating sympathetic dominance or a very slow recovery indicating parasympathetic dominance. Funkenstein (1956) extended the original observation and showed that three basic patterns were found in patients. Patients who could be classified as sympathetically hyporeactive showed prolonged depression of blood pressure following injection of the drug. Conversely, patients with hyperactive sympathetic nervous systems reacted to an injection by showing a prolonged rise in blood pressure. The test, sometimes called the Funkenstein test (Feinberg, 1958), is not used extensively today. However, the Mecholyl technique is important and it has been relied upon by Gellhorn to support many of his hypotheses concerning autonomic tuning (Gellhorn and Loofbourrow, 1963).

THE COLD PRESSOR TEST

The cold pressor test uses the normal physiological reactions of the autonomic nervous system to cold and heat. From Chapter 3 you will recall that cold produces vasoconstriction of the peripheral arterioles and heat produces vasodilation in the peripheral arterioles. In the cold pressor test you induce the sympathetic nervous system response of vasoconstriction to cold stress, measure the resultant increase in blood pressure and observe recovery of the blood pressure back to its prestress level.

The normal technique when applying the cold pressor test is to allow a subject to rest until his blood pressure reaches basal level. Then the arm, sometimes the foot, opposite to the one from which the blood pressure is being recorded is immersed into iced water for a period of one minute. Blood pressure readings are taken after the limb has been in the water for 30 seconds and again after one minute. The limb is then removed from the water and blood pressure readings are taken every two minutes until the blood pressure returns to its former base level. The highest reading recorded is taken as the cold pressor response (Hines and Brown, 1933). Lacey and Lacey (1962) gave a cold pressor test to a group of young children and retested them on a second cold pressor test four years later. The results of the second test revealed similar response patterns to the original test indicating individual response patterns that persist in spite of maturational changes that would have taken place during this period. Lovallo (1975) has reviewed the physiological and psychological research that has been carried out using the cold pressor test. He points out that the cold pressor response affects cortical and limbic structures via sensory input to the brain. Influences of higher brain areas are indicated by a study carried out by Igersheimer (1953) in which a cold pressor response decrement was shown in medical students who had been given an anaesthetic. The cold pressor response can also be modified by hypnosis and this is taken by Lavallo to indicate cortical involvement. Lavallo concludes that the cold pressor test is of value for obtaining a measure of autonomic balance in both normal and abnormal subjects.

THE COLD REWARMING TEST

This is a test of thermoregulation, the subject is placed in a room with an even temperature and an absence of air currents and the temperature recorded from one of his fingers until the readings are constant. The subject then immerses his hand and wrist for 10 minutes into water which has a temperature of 12-14 C. The hand is then removed and carefully dried without rubbing and the temperature from the same finger is measured every two minutes until it reaches the original value. Apart from the problem of producing vasodilation while drying the hand, other factors that can influence the results of this particular test are the position of the body and the hand. These should remain constant during the test. The initial skin temperature and the time that has elapsed since the last meal are also critical variables affecting this test.

DERMOGRAPHISM

In this test the inner side of the forearm is stroked with an object weighing about 150 grams and the time taken for the redness (erythema) to develop along the stroke mark is recorded. Three stages have been identified: a red weal which has a latency of about 15 seconds; a local flareup which develops over a period of 15 to 30 seconds; finally, the stage of urticaria which reaches its maximum after a period of five minutes. The dermographic response lasts for a period of up to 30 minutes. It is most pronounced in young children becoming slower and weaker in old age. It will be mentioned again in the next section as an important test when constructing an autonomic scale in young children.

A SCALE OF AUTONOMIC ACTIVITY

At a future period it is conceivable that it will become standard practice to carry out a series of autonomic tests, on individuals during their life-span, that will have the object of finding out not only the basal levels of individuals but also how an individual's autonomic nervous system reacts during stress. Lacey (1950) has tested invididuals over periods of time and concluded that organized patterns of autonomic reaction are shown in similar situations and that these patterns are repeatable over time.

It would clearly be of value to provide a scale of autonomic activity that would give a reliable measurement, not only of an individual's autonomic activity, but his autonomic activity relative to the total population. During this century there have been a number of attempts to provide such a scale and they have had varying degrees of success.

The first attempt using modern concepts of autonomic function was made by Eppinger and Hess (1915) who classified people on the basis of how their autonomic activity was divided between sympathetic and parasympathetic functions. Basically their idea was the concept of 'autonomic tuning' introduced in Chapter 3. The essential idea behind this concept was that the balance between the sympathetic and parasympathetic nervous systems reflected the type of personality shown. Eppinger and Hess spoke of a vagotonic or sympathotonic person and were concerned with the response of an individual's autonomic nervous systems to drugs. Their analysis was based upon responses shown to certain cholinergic and (nor)adrenergic drugs. The Eppinger and Hess classification never found complete favour; in part, this was because evidence was also beginning to accumulate to show that certain of the Eppinger and Hess assumptions, about very young children being vagotonic, were incorrect. In addition, there were other basic problems. For example, was the autonomic imbalance due to overactivity in one system or underactivity in the other? We shall return to consider this problem in more detail in the final chapter. However, it would be wrong to dismiss the concept of Eppinger and Hess entirely for, in many respects, it appears to have crucially influenced a number of later workers who include Gellhorn and Wenger.

M.A. Wenger has devoted his research career, spanning over 30 years, to the study of autonomic functions and part of this time has been spent attempting to devise a scale of autonomic activity (Wenger, 1941; Wenger, 1948; Wenger, 1966). From the results of his intercorrelational and factor analytical tests,

Wenger showed that seven peripheral autonomic cariables could give a profile of autonomic function. His autonomic measures were taken from resting subjects and are similar to the Eppinger and Hess test in that they are designed to measure the relative difference in function between the sympathetic and parasympathetic components of the autonomic nervous system. It should be noted that the scale is essentially a statistical one. Autonomic factors contributing to the scale for children were different from the factors found to be significant for adults, they were:

Children		Adults	
(1)	salivary output	(1)	salivary output
(2)	palmer conductance	(2)	palmer conductance
(3)	heart period	(3)	heart period
(4)	volar forearm conductance	(4)	volar forearm conductance
(5)	respiration	(5)	sweat response
(6)	pulse pressure	(6)	diastolic blood pressure
(7)	dermographic persistence	(7)	sublingual temperature

The reasons for the differences are not clear although some appear understandable. The dermographic test was mentioned earlier in this chapter and may relate to skin sensitivity, sweating might only become important when the surface area of the body becomes a critical size. In a child, the smaller body size results in a relatively greater body surface area enabling him to dissipate sufficient heat without the need for the additional mechanism of sweating. This temperature problem was mentioned in the last chapter when discussing temperature regulation in young babies.

The frequency distribution of mean estimates of autonomic balance $(\bar{A})$ for the seven autonomic measures in school children is given in Figure 4.1 taken from the Wenger *et al.* (1956). The mean score $(\bar{X})$ of the distribution is approximately 70 with a standard deviation of 8; thus, two-thirds of the distribution falls between the values 62 and 78. For any individual in the measured population, it is possible to determine the amount of deviation from the mean score and the direction of the deviation. Deviation from the mean value towards the left-hand side of the distribution shows relative sympathetic dominance (low $\bar{A}$ scores), deviation from the mean value towards the right-hand side shows relative parasympathetic dominance (high $\bar{A}$ scores). Of course, the scale only measures patterns of resting autonomic balance, patterns of reactivity might be totally different. It is not impossible that we could find a person who, although displaced to one side of the distribution normally, who might, during a stress situation, reverse completely into the opposite state. Despite certain differences, Wenger (1966) has claimed that later results have essentially confirmed the earlier studies. One crucial point concerns the reliability of the tests. It is a well known finding that psycho-physiological tests are not particularly reliable and

80

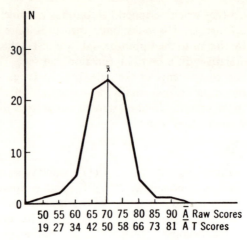

50 55 60 65 70 75 80 85 90 Ā Raw Scores
19 27 34 42 50 58 66 73 81 Ā T Scores

Figure 4.1 Wenger's distribution of autonomic balance scores. Using seven measures of autonomic function, Wenger plotted a frequency distribution of autonomic activity in a large number of subjects. The mean value was 70 with a standard deviation of 8; thus, two-thirds of the population fell between the values of 62 and 78. Deviations away from the mean value towards the left-hand side of the distribution represent relatively increasing levels of sympathetic dominance. Deviations away from the mean value towards the right-hand side of the distribution represent relatively increasing levels of parasympathetic dominance. Reproduced by arrangement with Holt, Rinehart & Winston, Inc., New York, from *Physiological Psychology* by Wenger, Jones and Jones, © 1956.

often difficult to repeat. Wenger (1966) has reported that the most reliable measure over time is salivary output which has given retest reliability coefficients of +0.88 to +0.79 over a three year period; however, some of the other tests give very low coefficients. Wenger feels that if extreme care is taken to duplicate the measurements, then they can be shown to be reliable over time.

One vital question relates to the validity of Wenger's autonomic scale, one might ask how valid is the $\bar{A}$ score as a general measure that can be applied outside of the limitations of the original sample. In the Presidential address to the Society for Psychophysiological Research, Wenger (1966) presented a number of findings that he suggested gave his scale more than limited validity. He quotes the research of McKilligott (1959) who tested paraplegics (these are patients who have suffered transection of their spinal cords resulting in reduced sympathetic activity) and found that their scores tended towards the parasympathetic end of the scale (low $\bar{A}$ scores). Hohmann (1966) divided this sample of paraplegics into groups with lesions at various levels and found correlations that indicated descending lesions or transections of the spinal cord were correlated with a progressive tendency toward the parasympathetic end of the scale (high $\bar{A}$ scores). In addition, students were tested shortly before an oral examination and were shown to have low $\bar{A}$ scores which were held to indicate sympathetic dominance and to reflect the anxiety they felt. Indeed, one student who was said to be very anxious dropped from his normal mean resting level of 78 down to 52. However, Gunderson (1953) and Sherry (1959) failed to find the predicted parasympathetic dominance for certain groups of schizophrenics and Sternbach (1966) has reported a low correlation of +0.18 between the electrical activity of the brain (alpha index) and the $\bar{A}$ score. Duffy (1972) has claimed that the autonomic scale may not be valid as it related to less than one-third of each population studied. Lacey (1950) has claimed that it is more valuable to record the individual patterns as these represent a more important and meaningful measure than data taken from groups of subjects.

Although these are clearly a number of difficulties, Wenger has carried out a valuable research programme, it is hoped that with the advent of sophisticated computers similar programmes employing longitudinal studies may be undertaken in the future.

THE LAW OF INITIAL VALUE

Psychophysiologists are concerned with amplifying minute electrical changes occurring within the body into quantifiable records. So far we have assumed that the physiological systems being recorded remain at a constant level but this is rarely, if ever, the case and it is this problem that we should consider next.

When considering biological systems we encounter organisms showing constantly fluctuating levels of their various internal processes. Cannon referred to these fluctuations as homeostasis, pointing out that they oscillate around a mean baseline level. These oscillations produce a problem in the stimulus/response (S-R) experimental situations used by psychologists and the problem concerns the point at which the stimulus is presented. Wilder (1957) proposed an empirical-statistical rule to deal with the problem of fluctuating biological rhythms which he called the Law of Initial Value. He later extended his idea to deal specifically with human bodily fuctions (Wilder, 1962). The Law of Initial Value predicts that the prestimulus level will affect the response and the basic problem is shown graphically in Figure 4.2, which shows a function that is oscillating in a biologically uncharacteristically regular manner between a maximum and a minimum value. Upon this ideal sinusoidal response curve, let us place three stimulus points of equal intensity: one occurring at a minimum value (S^1); the second one occurring halfway up the ascending slope (S^2); the third occurring at the maximum value (S^3). Because the response curve is bounded by a maximum and a minimum value, a stimulus point falling on the halfway point is unlikely to give a response (R^1) as large as the response (r) evoked by the stimulus given in the trough of the curve. Thus, the magnitude of

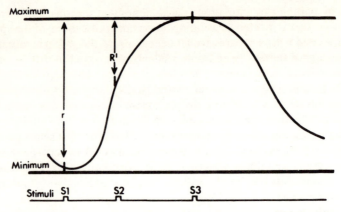

Figure 4.2 The Law of Initial Values. The sinusoidal function shown in this diagram represents a hypothetical internal autonomic function smoothly fluctuating between a maximum and a minimum value. Stimuli (S_1, S_2, S_3) of equal intensity and duration are presented at different points in time. Due to the independent oscillations of the function, S_1 occurs at a minimal value, S_2 at a mid value on the ascending slope, and S_3 at the maximum value of the function. The magnitude of the responses is different because of a timing relationship that is independent of the stimulus.

the evoked responses (R^1) and (r) differs because of a timing relationship that is independent of the stimulus. An actual example might serve to further clarify the significance of this law. Assume that we are measuring the autonomic function of heart-rate and, moreover, we can consider it as an isolated autonomic process of increasing beat controlled by the sympathetic nervous system and decreasing beat controlled by the parasympathetic nervous system. In this oversimplified example we have a basic homeostatic process in which, if we consider it as a cybernetic model, any increase in heart-rate will tend to return to the baseline value by a process of negative feedback. The feedback is positive, meaning that the higher the heart-rate the greater will be the forces attempting to return it to the resting value. Any change of the heart-rate away from its resting level will be opposed by physiological mechanisms attempting to restore it to its resting level. An increase of from 66 to 76 beats per minute is not equivalent to an increase of 86 to 96 beats per minute as the inhibiting forces at the higher values will be greater. Ideally, we need a correction factor that would make stimuli equivalent wherever they fell on the fluctuating curve. Even in this simple theoretical model, the situation worsens if the stimulus coincides with the maximum value of the slope (S^3). In this event, we might find a decrease in heart-rate when our hypothesis led us to predict an increase in rate.

If we consider autonomic functions as they are normally recorded then the situation is infinitely more complex, not only do we find complex interlocking loops and systems but the fluctuating level of activity is nothing like the smooth sinusoidal function described above. So it would seem to be dubious practice to

rather than the quantity of activity. For example, during sleep activity of the brain, as measured by the electroencephalogram, shows decreased electrical activity but we find increased activity in certain parts of the autonomic nervous system. Indeed, if giving the autonomic nervous system adequate recovery time is not one of the main functions of sleep then it appears that such a result is a fortuitous consequence of sleep. However, before we look specifically at the correlation between the brain and the autonomic nervous system we will briefly consider the history of the concept of arousal.

The two most famous physiological concepts of arousal have both been encountered in this book. The first was Cannon's 'fight or flight' role for the sympathetic nervous system. The second was Pavlov's orienting response. Both have been broken down or fractionated into smaller components. The components of the orienting response have been mentioned in the last section and the major attempt to analyse the component parts of the sympathetic response will be considered in the next section when we look at the work of John Lacey. Note, that as originally stated, both were concepts of general arousal. This is also true of Duffy (1934) who introduced the term as a psychological concept at a time when growing interest in the newly developing field of electro-encephalography made the theory plausible for many psychologists who saw a possible physiological base for the theory. The concept was helped by the discovery of Magoun and his colleagues (Lindsley et al., 1949; Moruzzi and Magoun, 1949; Lindsley et al., 1950) that the midbrain reticular formation seemed to serve as a general arousing mechanism for the cortex. Lindsley (1951) proposed a specific theory of activation in which the cortex was aroused by electrical discharges from the reticular formation. The original theory has been extended by Lindsley (1970) to give a larger role to the thalamus and to emphasize relaying of information back from the cortex to the reticular formation. As originally proposed the theory argued that the physiological basis of motivation was reducible to the idea of the reticular formation maintaining a critical level of activation in the cortex. Routtenberg (1968) has proposed a more sophisticated two-stage hypothesis in which the limbic and the reticular activating systems are stated to have separate activating roles.

The reticular activating system is not a simple homogeneous brain structure but complex aggregations of nuclei and neuronal pathways that can be shown to have specific effects by using electrical, surgical, and pharmacological techniques to unravel the various systems. Glickman and Schiff (1967) have reviewed evidence that points to different but specific arousal mechanisms in many species. Grossman (1973) presents further evidence against the idea that the reticular activating system is the sole arousing mechanism in the brain and his chapter on this topic forms a very good introduction to the discussion of general versus specific motivational theorists.

To return to the original concept proposed by Duffy (1934) her concept of arousal was initially proposed to explain variations in responsiveness to stimulation that were not related to changes in specific drive states; it referred to the energy manifested in the underlying physiological processes. Over the years the theory has been modified considerably to take account of the objects and

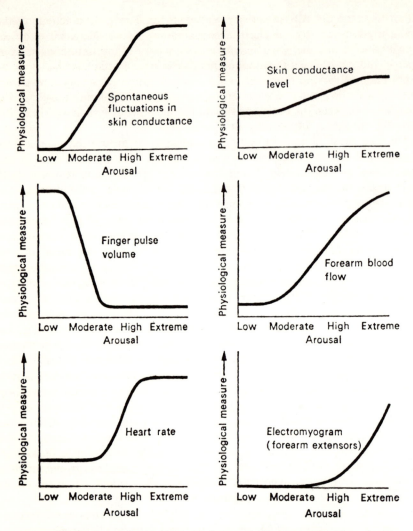

Figure 4.3 Physiological indicators of arousal. The parameters from low to extreme arousal are shown for five psychophysiological systems. Note that in the case of finger pulse volume the more normal increase in level is reversed and a decrease is shown in this measure at the higher levels of arousal. After Lader (1975), *The Psychophysiology of Mental Illness*, Routledge & Kegan Paul Ltd, London.

difficulties of proving any underlying physiological process that governed the overall level of arousal. The major weakness of the theory is that it is usually concerned with a single dimension that may or may not be important in the behaviour being studied. Overall, most of the workers who have attempted to study the underlying physiological processes would have been better advised to look at the finer details of the systems they were using rather than lumping them together in a grand manner. This is one of the major criticisms made by Duffy (1972) in an important review paper written shortly before her death.

Duffy points out that we have little knowledge about the best way of combining different physiological measures that may well have quite different time courses. Lader (1975) has presented a series of schematic curves showing various physiological measures and they are reproduced in Figure 4.3. The different measures are plotted along the continuum from low to extreme arousal. In the curve relating to finger pulse volume it can be seen that the usual increase in level of activity from low to high arousal is reversed and contains a very sharp drop. Muscular activity shows very little activity until the level of arousal reaches fairly high levels. Duffy feels that in many cases the underlying arousal mechanisms would be shown if the investigators had used the maximum or peak level reached. However, it is difficult to suppose that this measure would not have been used given the general tendency of psychophysiologists to transform their data to obtain significant statistical levels. Duffy admits that the support for the concept of arousal is equivocal. In a very detailed study Sternbach (1960) used 11 measures of autonomic activity and a number of stimulus conditions that included startle, cold pressor test, and exercise as well as injections of adrenaline and noradrenaline but found only low negative correlations. Elliott (1964) analysed the correlations between autonomic and somatic measures in children and young adults and concluded that the low negative correlations that he found suggested that there was not a common arousing mechanism in the body. Given the situation outline in Figure 4.3 it is scarcely surprising to find a lack of correlation in these studies.

The main problem that appears to have arisen from the use of the concept of arousal is that it has been used too loosely by psychologists with limited knowledge about the underlying physiological processes. Broadbent (1977) when discussing arousal says 'alcohol raises level of arousal while it is known from physiological evidence that it is a depressant rather than a stimulant'. If we think about this problem we conclude that there is no reason why alcohol at different concentrations should not be both a stimulant and a depressant; indeed, this is what its physiological action appears to be. The term arousal with its apparent elasticity has often seemed to be the 'Polyfiller' of psychology used to fill any gap or crack in knowledge. Discoveries and benefits have undoubtedly arisen from attempts to prove or disprove the concept of arousal, but on balance it seems that this effort would have been better spent in attempting to understand the details of the underlying physiological mechanisms.

FRACTIONATION OF AUTONOMIC FUNCTION

Ax (1953) suggested that if correlations of sympathetic and parasympathetic function were considered, it was found that there was a consistently larger correlation for the state of anger than the state of fear. He argued that during the course of evolution a successful attack required greater organization of the bodily functions than flight. Later he elaborated these ideas and suggested that fear and anger could be accurately labelled using peripheral measures of autonomic activity.

Lacey (1967) went beyond this idea and suggested that it was possible to fractionate autonomic functions into perceptual, sensorimotor, and cognitive components. Lacey's theory arose from the finding that subjects placed into an experimental situation requiring them to detect a visual or auditory stimulus typically showed a decrease in their heartrates. Alternatively, if the subjects were placed in an experimental situation requiring them to reject the environment in favour of cognition, that is, carry out mental arithmetic tasks, they show increases in their heart-rates. From the evolutionary point of view, this is a surprising result. In the signal detection situation Cannon's activation theory, for example, would predict an increase in heart-rate in order to prepare the animal for possible avoiding action. Conversely, thought processes requiring rejection of most, if not all, of the environment should result in the heart slowing down as there would be little need for body activation. Intuitively it does not seem likely that when we fall into meditation or deep thought, we are particularly prepared for environmental emergencies; indeed, it is often very difficult to arouse a person who is deep in contemplation. Lacey also showed that the expectation of a painful stimulus causes a decrease in heart-rate. This, again, is in contradiction to the position taken by Cannon who argued that in states of extreme fear, the sympathetic nervous system became highly active.

Heart-rate decreases in response to fearful situations are also supported by Gellhorn (1965) who has pointed out that in states of extreme fear, animals die a 'vagal death' where the heart-rate slows down until it finally stops altogether. The phenomenon of the 'vagal death' was discovered by Richter (1957) who forced wild rats to swim for survival and discovered that if he removed the vibrissae or snout whiskers from the animals that they dived to the bottom of his tank and died almost immediately. Binik *et al.* (1977) have largely confirmed the original findings of Richter and extended them to the laboratory rat.

As Lacey has recorded, earlier investigators had reported similar changes in heart-rate, but these observations had never been followed up. Darrow *et al.* (1942) reported that sensory and ideational stimuli produced opposite effects on cardiac-rate. Sensory stimuli requiring no extensive association of ideas resulted in a decrease of heartrate, while noxious stimuli or sequences of activity requiring association produced increases in heart-rate. Lacey has suggested the idea of a continuum, with at one end stimuli that require momentary attention characteristically these result in fast reaction times. At the other end of the continuum are stimuli that involve cognitive organization, characteristically these result in slow reaction times. An example of the former would be detection of a flash of light and an example of the latter would be mental arithmetic. At the momentary attention end of the continuum we find a fall in heart-rate and blood pressure. At the opposite end of the continuum where cognitive activity is required, heart-rate and blood pressure show increases.

Lacey's most radical proposal has been to appear to suggest that the cardiac system was instrumental in controlling the activity of the brain via the pressure-sensitive receptors of the carotid and baroreceptors of the aorta. However, Malmo (1972) has argued that very careful reading of Lacey's position reveals that heart-rate slowing is taken merely as a sign that the corotid sinus mechanism is inoperative at that time. (The carotid sinus reflex is part of a mechanism that normally prevents the blood pressure from remaining at high levels. See, Vander *et al.*, 1970, for a very clear account of the function and structure of the cardiac system.) The idea that autonomic function may influence the brain was introduced in Chapter 3 (Kempf), but it might appear that Lacey has made the idea more persuasive by presenting experimental evidence. Searching the literature for further evidence to support Lacey's position reveals little, although it has been claimed that dreaming only takes place during activity of the sympathetic nervous system (Hersch *et al.*, 1970). This finding suggests a link between cognition, which is what dreaming presumably is, and the sympathetic nervous sytem.

Lacey's original concept has become less clear since its original formulation and, as Hahn (1973) has said, the concept of attention, never rigorously defined by Lacey, has not been helped by the use of phrases such as; 'the subject wants to accept' or 'wants to reject'. Other investigators have suggested that insufficient attention has been given to the fact that it is possible that subjects are using subvocal verbalization and this is an important confounding variable in the type of experimental situation used by Lacey (Campos and Johnson, 1967). Others (Webb and Obrist, 1970) have reported that muscle activity (EMG) and heart-rate show a negative correlation with reaction time. From this they conclude that cardiac changes occur as part of a general somatic pattern which theoretically is in accord with the Russian position.

The heart-rate changes first demonstrated by Lacey have been reported by others but it is doubtful if they do demonstrate autonomic control of brain function. Lacey (1967) has evaluated the neurophysiological evidence for and against his position but he has largely ignored the peripheral factors that serve to control heart-rate. The interested reader will find a review of the peripheral factors in Burnstock (1969). The heart, as its crucial role in maintaining the bodily function dictates, sits among many controlling loops that all serve to influence its activity at one or other point. It seems dubious, given our current knowledge, to select one point from these loops and claim that it is the crucial point. All are equally important, evaluation of the environment must be made in the brain by both nervous and hormonal means. This evaluation, will alter the heart-rate that, in turn, will feed back and modify aspects of brain function. If we consider the vascular system which is largely under control of the sympathetic nervous system, we find that the volume of blood is considerably smaller than the total area of the blood vessels, and large redistributions of the blood, that are not reflected in a change of heart-rate, may take place by vasoconstriction.

At the present time, it seems better to favour a somatic inhibitory patterning in which the blood flow and rate reflect decisions by the brain and serve to prepare muscles accordingly. However, until we can reliably measure blood pressure and stroke volumes of humans (the problem arises from the need to insert catheters into blood vessels to obtain these measures), it will be difficult to prove Lacey right or wrong. As Darrow (1943) pointed out many years ago, understanding of cardiac function must include consideration of the inter-relationship between blood pressure and heart-rate. But for all these critical points, Lacey's suggestion has attracted wide attention and more than any other person, he was responsible for providing evidence to break away from Cannon's idea of an emergency function for the sympathetic nervous system. His idea of autonomic fractionation gave a valuable new perspective to the workers in psychophysiology and stimulated much needed and valuable research.

In large part Lacey's careful work helped to pave the way for the emphasis in psychophysiology to switch from studies concerned with the principles of behaviourism and learning theory to studies concerned with clinical problems and attempting to understand the physiological systems they were working with in terms of their possible biological significance. The importance of this latter point has been made by Obrist (1976) who pointed to the importance for the need for greater biological sophistication in psychophysiology and for the realiz-ation that the biological properties of the systems being studied are important.

Obrist (1976) has summarized his careful studies of the cardiovascular system and come to the following conclusions. In experimental situations where the subject is a helpless recipient of an aversive event, the heart is primarily under vagal control and blood pressure shows modest phasic changes which are mainly vasoconstrictive; somatically the subject becomes momentarily immobilized. In experimental situations where the subject is able to cope, but not totally, very large increases in heart-rate are found which in 25 per cent of the subjects are reported to be in excess of 40 beats per minute, this indicates sympathetic control Systolic blood pressure is under cardiac control with diastolic blood pressure showing little alteration indicating influence of the sympathetic system. For a complete overview of the work being carried out in the area of cardiovascular research the interested reader is referred to Obrist *et al.* (1974) where he will find a series of review papers.

FUTURE TRENDS IN PSYCHOPHYSIOLOGY

The early emphasis psychophysiologists placed upon autonomic function is not difficult to understand, the autonomic system is readily accessible and seemingly has discrete functions. Thus, when the 'Zeitgeist' required psychological research to be limited to publically observable functions the invention of the polygraph produced an acceptable instrument. At first, the tracings of the bioelectrical potentials produced by the underlying physiological activity were traced onto a continuous roll of paper and analysed by qualitative methods, but an increasingly sophisticated technology now not made it possible to produce instantaneous data analysis using computers. The usual statistical treatment of psychophysiological data has normally involved correlational techniques but Ax (1964) has argued that psychophysiologists must advance beyond correlational techniques and attempt to translate the psychological-physiological code in order to obtain a more complete understanding of the underlying processes. This problem would involve more extensive monitoring of the particular autonomic system being studied. For example, if we take skin temperature, it is not sufficient to measure only one aspect. Two subjects placed in a warm environment might display similar skin temperatures but be using separate mechanisms of their sympathetic nervous systems to achieve this similar temperature. One subject may compensate for the environmental temperature by increased sweating, while the other might achieve the same result by virtue of more efficient control over vasodilation of his peripheral blood vessels. In most cases it is a gross oversimplification to use a discrete measure of autonomic function. This problem is exemplified by the use which has been made of the galvanic skin response (GSR) as a discrete measure. The GSR is dependent upon a number of bodily systems (Malmo, 1965) but a number of ingenious statistical treatments have been used to demonstrate that the galvanic skin response was a reliable and meaningful measure, e.g.:

(1) logarithm of the change in conductance;
(2) the change in the logarithm of conductance;
(3) logarithm of the change in resistance;
(4) the ratio of the logarithm of the change plus a constant all over the logarithm of the initial resting level;
(5) percentage change in resistance;
(6) square root of conductance;
(7) percentage of ohmic decrease of the maximum decrease obtained.

The transformations are not rooted in physiological function but have emerged as the result of statistical manipulations. They often turn out to have a specific relevance to the original data and are, therefore, of little use when making comparisons between different studies. Given our present psychophysiology knowledge we must be cautious when attempting to use discrete measures of autonomic function as indices of general psychological concepts.

In a Presidential address that looked into possible future developments in psychophysiology, Brown (1966) has suggested that, apart from the usual techniques, attempts must be made to develop new ones. One interesting possibility, mentioned by Brown, is the work by Becker *et al.* (1962) who suggested that the slow fluctuating direct current (d.c.) voltages that can be measured upon the surface of the skin, exert control over the body that is supplementary to the nervous system control as we understand it at the current time. The d.c. voltages might control speed of neuronal impulses and hence mood and feeling. A recent trend in psychophysiology concerns attempts to measure cognitive aspects of brain function (McGuigan and Schoonover, 1973). This type of development should be encouraged and extended in an attempt to understand exactly how autonomic and cognitive functions intermesh with each other. Another developing area with psychophysiology is the study of the autonomic processes during states of meditation (Tart, 1969; Orme-Johnson, 1973). Wallace *et al.* (1971) have shown that during a meditation state, decreased oxygen consumption and carbon dioxide expiration rate indicated a hypometabolic state. Other workers have not been so optimistic and have concluded that activity of the brain and autonomic nervous system cannot be used to define states of consciousness (Johnson, 1970). At the present time we should be cautious in interpreting studies relating to meditation. Biorhythms is another area which promises many interesting findings and reports are beginning to appear. In recent years there has been a resurgence of interest in control of autonomic functions by learning processes called biofeedback, but this area has grown so large that it warrants separate discussion in the next chapter. Hopefully one area of growth will be the use of computers to make detailed longitudinal studies on the autonomic activity of individuals.

THE PSYCHOPHARMACOLOGY OF THE AUTONOMIC NERVOUS SYSTEM

Individual nerve cells operate by complex series of electrochemical events within the cell. During the past two decades we have extended our conception of brain function by beginning to understand how drugs alter the chemical pathways of the brain. The chemical codings of the brain can now be seen to be as important as the electrical activity in the brain. In this section we will largely confine discussion of psychopharmacology to drugs that have a specific effect on the autonomic nervous system. We will also discuss certain aspects of the chemistry of the autonomic nervous system which have been shown to have behavioural effects. A comprehensive discussion of psychophysiology and drugs will be found in Stroebel (1972) who has an excellent section on the autonomic nervous system. Warburton (1975) discusses drug effects on brain function and gives a most complete discussion of the effects of drugs on the cholinergic pathways in the brain. For a detailed overview the reader is referred to the latest edition of Goodman and Gilman (1970).

As pointed out at a number of places in this book the science of pharmacology played important roles in the discoveries and understanding of the structure and function of the autonomic nervous system. The use, by Langley, of iodine to paint individual ganglia of the sympathetic nervous system enabled the system to be mapped early in this century to determine its structure. Similarly the Mecholyl test is an example of a drug used to examine the function of the autonomic nervous system within individuals. Alquist (1948) clarified a number of the paradoxes concerning the function of the autonomic nervous system with his concept of receptor membranes of (nor)adrenergic nerve cells having alpha- and beta-receptors. More recently, as indicated in Chapter 3, the importance of the (nor)adrenergic pathways in the brain has begun to be understood and our knowledge of how these pathways can be modified by the action of drugs is increasing rapidly.

The popular history of drugs appears to fall into three stages; vitamins, antibiotics, and tranquillizers. The latter compounds are now prescribed in ever increasing numbers and this trend has come to influence our conceptions about both the brain and the autonomic nervous system. This influence will be seen in Chapter 5 when biofeedback is discussed. Scientific interest in drugs and their behavioural effects began in the 1940s and 1950s with studies related to the actions of acetylcholine. During the mid-1950s the biogenic substance serotonin came briefly to the forefront because of its presumed relationship to the hallucinatory drug lysergic acid. Interest then switched to the biogenic transmitter substances involved in the (nor)adrenergic pathways of the brain, which finally became epitomized in the catecholamine hypothesis of affective states which stated that mood was directly related to the amount of catecholamine substances in the brain at any time. This will be discussed later; but at this point it is sufficient to say that the case was somewhat overstated. During the 1970s interest is changing away from the role of noradrenaline to considerations about the role of dopamine (Iversen, 1975), and the mechanism by which catecholamines come to influence sources of metabolic energy used by cells. Cells store energy in the form of animal starch in which the energy is taken up and released by the adenosine monophosphate and triphosphate cycle which has been called the secret of life.

Table 4.1 Table showing some of the drugs used to block or mimic autonomic and somatic receptors.

	Preganglionic	Postganglionic
Sympathetic nervous system		
Neurotransmitter substance	Acetylcholine	Noradrenaline
Mimetic drug	Dimethylphenyl-piperazinium (DMPP)	Phenylephrine
Blocking drug	Hexamethonium	
alpha-receptors		Phenoxybenzamine
beta-receptors		Propranolol
Parasympathetic nervous system and (sympathetic sweat glands)		
Neurotransmitter substance	Acetylcholine	Acetylcholine
Mimetic drug	Dimethylphenyl-piperzinium (DMPP)	Muscarine
Blocking drug	Hexamethonium	Atropine
Somatic nervous system		
Neurotransmitter substance	Acetylcholine	Acetylcholine
Mimetic drug	Tremorine	Phenyltrimethyl-ammonium (PTMA)
Blocking drug	Atropine	Decamethonium

Table 4.1 shows the transmitter substances and an example of a major mimetic drug and a major blocking drug for the peripheral nervous system. Note, as shown in Figure 2.2 that the preganglionic transmitter substance for the sympathetic nervous system is acetylcholine and the postganglionic transmitter substance is noradrenaline. When considering blocking drugs affecting the sympathetic nervous system, it must be remembered that it is possible to block alpha- and beta-receptors selectively and many drugs are produced with the aim of blocking one or the other. The pharmacological properties of the sympathetic nervous system are complicated by the fact that the sweat glands of the body are mediated postganglionically by the neurotransmitter substance acetylcholine. The sympathetic nerves to the adrenal medullae are also preganglionic, so it is not surprising to find their effects mediated by acetylcholine. The drugs listed in Table 4.1 can be used to control the amount of available transmitter substance or to excite or depress the receptors upon the surfaces of the receptive membrane.

Longo (1972) has presented a schematic classification of drugs in terms of sympathomimetic, sympatholoytic, and parasympatholytic effects. Longo also includes a classification of drugs he calls 'no autonomic effects' but it must be remembered that all substances in sufficient concentration will show pharmacological effects. For example, the inclusion of alcohol as having 'no effect' can only mean at moderate concentrations; a 'hangover' reveals plenty of autonomic effects to alcohol. Account of the action of drugs on the autonomic nervous system can be found in Jenkinson (1973) and Johnson and Spalding (1974). For more detailed consideration of the action of drugs the following references will be valuable sources (Goth, 1974; Goldstein *et al.* 1974).

A schematic drawing of a (nor)adrenergic nerve cell is shown in Figure 4.4. A number of drugs that affect the various metabolic stages outlined below are shown in the figure and their main effects in terms of potentiation or inhibition are indicated in the figure by a plus or minus sign alongside each drug. Five main stages are shown in the schematic figure, they are: synthesis, storage, release from the nerve cell, activity at the receptor surface, and finally, the fate of the released noradrenaline.

The metabolic steps in the synthesis are shown in the small inset in the figure. The synthesis of noradrenaline is thought of as a continuous process and this process may be blocked at the various stages of metabolism. The drug alpha methyl tyrosine is given as an example of a drug that has a metabolic blocking action. After it has been synthesized noradrenaline is actively transported into vesicles where it appears to be stored in two forms. The looser bound form, indicated by the dotted circle around it, is sensitive to depletion by the drug tyramine. The more resistant form, indicated by the continuous circle, is not sensitive to depletion by tyramine (Iversen, 1967). Both the free and the stored noradrenaline may be depleted by monoamine oxidase (MAO) and certain other drugs, e.g., reserpine. Drugs that serve to inhibit or deaminate the transmitter substance reduce the amount of noradrenaline at the end of the

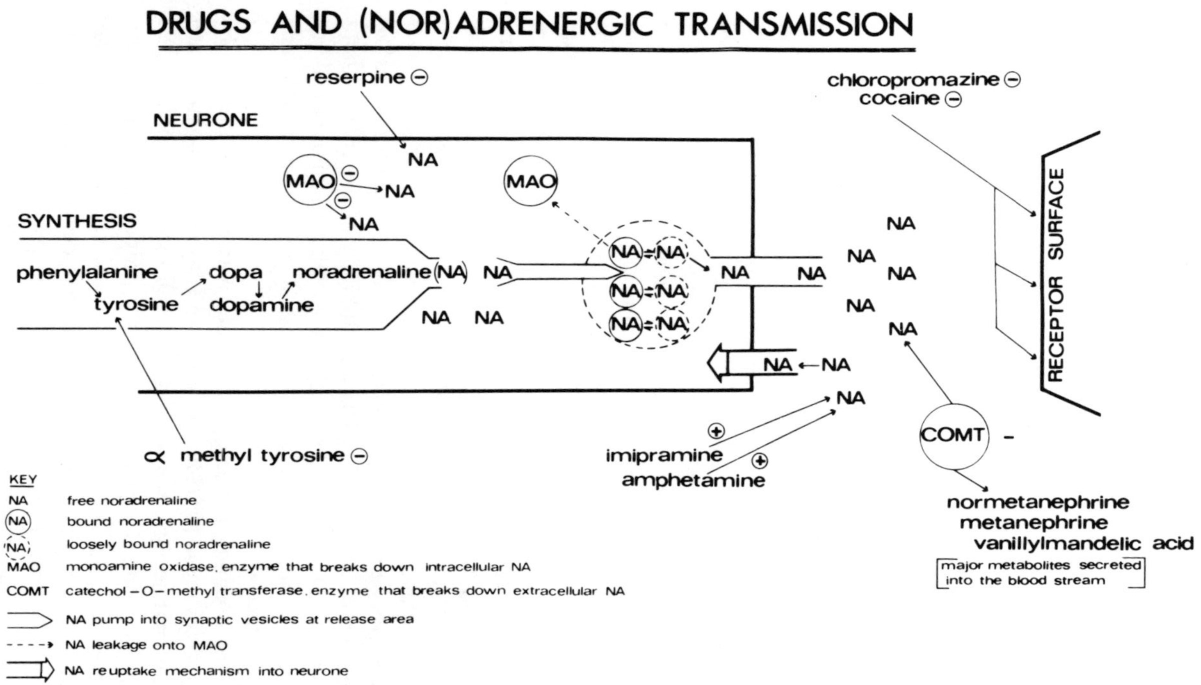

DRUGS AND (NOR)ADRENERGIC TRANSMISSION

Figure 4.4 The pharmacology of a (nor)adrenergic nerve. The schematic diagram shows the end of a sympathetic nerve together with a synaptic cleft. The key at the bottom left of the diagram explains the major aspects of biosynthesis, storage, and release of noradrenaline (NA) shown in the diagram. Following biosynthesis, NA is taken up into a bound form and stored until released as a transmitter. Release of NA stimulates the receptor surface of an adjacent neurone. As indicated in the diagram, extraneuronal NA is either broken down into metabolites that are excreted in the urine or actively retaken back into the neurone. By selective use of drugs it is possible to interfere with NA at various stages of biosynthesis, storage and release. Based on a drawing by Dr. Joseph L. Slangen.

nerve cell. The action of MAO in inhibiting noradrenaline may be blocked by MAO inhibitors such as iproniazid or tranylcypromine. The arrival of an action potential at the end of a (nor)adrenergic axon results in the release of noradrenaline into the synaptic cleft in relatively large amounts. As indicated in Figure 4.4 a certain level of the transmitter substance will always be present at the nerve ending. It should also be realized that the action potential itself may be enhanced or blocked; for example, a drug that enhances an action potential is guanthidine and a drug that blocks an action potential is chlorisondamine. Once released into the synaptic cleft the transmitter substance may follow several pathways but for the moment we will consider the action of noradrenaline that reaches the receptor surface of the adjacent nerve cell. Noradrenaline reaching the receptor surface will produce a graded depolarization that spreads possively through the dendritic processes of the nerve cell into the cell body where, at the axon hillock, it may join with other arriving impulses to initiate or inhibit an action potential in the cell.

From previous discussion it will be clear that the receptor surface of a (nor)adrenergic nerve cell may contain alpha- or beta-receptors. Alpha-receptor sites, associated with vasoconstriction, may be blocked with the drugs chlorpromazine, phentolamine or phenoxybenzamine. Beta-receptor sites, associated with vasodilation, can be blocked by the drug propranolol. Many pharmacological compounds are now produced with the aim of having a specific action on one or other of the (nor)adrenergic receptor sites. Similarly, adrenaline-like and noradrenaline-like compounds may be used. For example, if we consider the action of adrenaline and noradrenaline on smooth muscle we find that adrenaline has both an excitatory (motor) action and an inhibitory action, whereas noradrenaline has only an excitatory action. Treatment of asthma requires a pharmacological compound that will relax the muscles of the bronchioles. Because of its basic lack of inhibitory properties, noradrenaline is of very little use. The inhibitory action of adrenaline is about 40 times stronger than noradrenaline. However, a drug synthesized in the laboratory called isoprenaline (isoproternol in the USA) has an inhibitory action that is about nine times stronger than adrenaline. This ability to relax the muscles of the bronchioles means that isoprenaline is often used in preference to naturally occurring catecholamines.

Noradrenaline not reaching the receptor surface may be broken down by catechol-O-methyl transferase (COMT) which is an enzyme that breaks down extracellular noradrenaline into metabolites that will eventually be excreted from the body in the urine. Urine also contains a proportion of free noradrenaline and adrenaline. It is known that MAO and COMT are not very efficient inhibitors of noradrenaline and this finding has led to a suggestion that they may only operate when there is an excess of intra or extracellular noradrenaline. Extracellular noradrenaline may be potentiated by the drugs imipramine and amphetamine. Amphetamine can actively cause noradrenaline to be removed from the (nor)adrenergic cell to maintain short-term high levels of extracellular noradrenaline with its socalled 'speed' effects. This short-term activation eventually rebounds into behavioural fatigue because noradrenaline is removed from the nerve cell faster than it can be metabolized.

The intensity of metabolic activity within a (nor)adrenergic nerve cell may be gauged from the findings of Sjarne *et al.* (1969) who, using observations taken from the muscles of cats, calculated that the total amount of neurotransmitter substance stored inside (nor)adrenergic nerve cell was sufficient for 10,000 impulses. However, an average nerve cell may release up to five times this quantity of transmitter substance in one hour.

The first psychopharmacological concept that had a marked, if not a dramatic, effect on the idea that the autonomic nervous system determined personality was made by Eppinger and Hess (1915). These authors suggested that personality could be revealed by reactions to drugs having a specific effect on the sympathetic or the parasympathetic nervous systems. A person with a dominant sympathetic nervous system would show pronounced reactions to sympathomimetic and sympatholytic drugs and less pronounced reactions to parasympathomimetic and parasympatholytic drugs. A person with a dominant parasympathetic nervous system would show pronounced reactions to parasympathomimetic and parasympatholytic drugs and less pronounced effects to sympathomimetic and sympatholytic drugs. Darrow (1943) pointed out that one of the major problems with this concept was that it was difficult to decide if the reaction to a drug was due to an increase in the activity of one of the branches of the autonomic nervous system or a reduced activity in the other branch. We shall return to reconsider this problem when we deal with the problem of autonomic reaction in old people. Also, certain individuals display increased activity in both branches of their autonomic system and display exaggerated response to all four classes of the drugs indicated above. In addition it is a well known fact that the baseline of autonomic activity is affected by the administration of drugs and the technique used by Eppinger and Hess relied upon the administration of the drugs before measurements were taken. One type of research that arose from the Eppinger and Hess concept was the idea of an autonomic scale used by Wenger and the use of the Mecholyl test to determine sympathetic/parasympathetic levels. Both of these techniques were considered in the section dealing with psychophysiology and there is little need to do more than draw attention to them here.

CATECHOLAMINES AND BEHAVIOUR

The remainder of this chapter will consider the behavioural effects of the autonomic nervous system's natural neurotransmitters. Two main techniques have been used. The first is to obtain, by natural or experimental means, a behavioural change and then to examine vascular or urinary excretion rates of the catecholamines and their breakdown products or metabolites. The second technique involves injecting separately, or as a mixture, noradrenaline and adrenaline.

What exactly is the relationship between the catecholamines and behaviour?

Goodall (1951) claimed that the level of circulating catecholamines was related to certain behaviour characteristics. Aggressive animals, such as lions, had a higher proportion of noradrenaline in their adrenal glands while timid animals, such as rabbits, had relatively higher levels of adrenaline. This suggestion, however poetic, has not subsequently been confirmed. It was, however, of interest as it appeared to form a link with the psychological findings of Ax (1953) and Funkenstein (1956) who argued that fear and anger states in humans are characterized by a differential patterning of the autonomic nervous system. Anger was said to produce a noradrenaline-like reaction (Funkenstein) or a mixed adrenaline/noradrenaline-like reaction (Ax). It was agreed by both these investigators that fear produced an adrenaline-like patterning. Von Euler (1964) has discussed the possibility of quantifying stress by saying that mental stress involving exhilarating or aggressive reactions was associated with increased levels of circulating noradrenaline. Mental stress involving apprehension, anxiety, pain, or general discomfort was associated with increased levels of circulating adrenaline.

Goodall (1962) assayed urinary samples taken from subjects exposed to gravitational stress by use of a large centrifuge; he found that the adrenaline levels were related to the amount of anxiety evoked by the situation. Noradrenaline was related to physiological stress induced by being whirled around in the centrifuge. In general, Goodall found that adrenaline levels decreased gradually over days as a subject's anxiety level adapted to the psychological stress involved in the centrifugation process. The physiological stress induced by the centrifugation process remained constant over days, and this was reflected by noradrenaline levels not showing the adaptations found in adrenaline. Subjects showing high anxiety levels in the situation did not show progressive decrements in their excreted adrenaline levels.

Levi (1965) measured the excreted levels of catecholamines in urinary samples taken from women viewing films that were classified as emotionally neutral, emotionally pleasant, or emotionally unpleasant. He reported increased levels of both adrenaline and noradrenaline associated with pleasant films. Unpleasant films resulted in increased levels of noradrenaline alone. The levels of the excreted catecholamines appeared to be releated to the intensity rather than the type of emotion induced by the films. This point will be discussed again when emotion is considered.

Brady (1967) reported that in the monkey, both catecholamine levels increased when the animals were placed in situations characterized by novelty and uncertainty. When the animals were placed in familiar situations only noradrenaline levels increased. Brady also showed that increases in noradrenaline were found in situations where aversive stimulation was used. To explain these findings, Mandler (1967) suggested an explanation in terms of 'response availability'. He argued that adrenaline levels increase in situations where the animal does not 'know' what the correct behavioural response is. In situations where the animal does 'know' the correct response, increased noradrenaline is found. For example, if an animal is being shocked, then it will make an escape or an avoidance reaction. However futile it may be in fact, the animal has a defined behavioural response. This suggests that noradrenaline levels may somehow relate to the 'muscular activity' of the situation. It may be that the 'muscular activity' need not be overt and that subvert central nervous system activity results in the release of noradrenaline. This suggests that noradrenaline levels may relate to the sensorimotor processes of the extrapyramidal system or via other central motor processes in an, as yet, unspecified relationship.

It should be remembered that all the studies reported in this chapter refer to the relative levels or amounts of the two catecholamines in circulation. At any one time adrenaline will always be present in the largest amount. Also this area of research is bedevilled by conditioning complications. Release of catecholamines from the adrenals is accompanied by certain patterns of behaviour and feelings and these, through conditioning and learning, may come to accentuate if not actually produce correlations between behaviour and catecholamine excretion. Before the final relationships between mood and behaviour are comprehended, conditioning factors will need to be understood in some detail.

ATTEMPTS TO INDUCE MOOD CHANGES BY INJECTIONS OF ADRENALINE AND NORADRENALINE

Injection of adrenaline and noradrenaline into human subjects to determine if mood changes can be induced has a long and confused history that dates back to Marañon (1924). So many different levels of injections and infusion rates have been used that it is difficult to generalize from one experiment to another. At least part of the earlier confusion arose from the use of crude adrenal extract that contained a mixture of adrenaline and noradrenaline. However, it is true to say that the more recent studies using pure adrenaline have not revealed a picture that is much clearer. In general, subjects who received injections of adrenaline are reported as showing 'cold emotion'; that is, they have vague feelings of apprehension and anxiety but they are unable to fix their feelings accurately. It is as if their feelings were free-floating. It is also reasonable to suppose that a more discerning subject will hesitate to report feelings in the absence of an identifying cause. For example, anxiety often has a cause that can be recognized by the subject. This, like the problem of previous conditioning and learning pointed out above, will also tend to increase the variance shown both in catecholamine levels and subjective reports.

Schachter and Singer (1962) have elucidated some aspects of this problem by suggesting that psychosocial cues are important. They used a technique that involved injecting their subjects with adrenaline and placing them in situations where clearcut feelings and emotions were being displayed by another 'subject' acting out a prescribed role. They found that their psychosocial situations tended to polarize their injected subject's emotions and feelings which resulted in their subjects acting in a similar manner to the 'stooge' subject. In this connection, it is interesting to note that some of the earlier reports of unambiguous feelings induced by injection of adrenaline were made by psychiatrists using themselves as subjects or else using subjects in a psychiatric setting surrounded by doctors. Schachter and Singer's experiments have been criticized (Plutchik and Ax, 1967), but they are, nevertheless, very interesting studies that deserve to be repeated under more rigorously controlled conditions. They will be discussed again when we consider emotion.

Frankenhaeuser (1971) has summarized a series of careful experiments carried out with collaborators in which adrenaline, noradrenaline, and a mixture of the two were infused into subjects over long periods. She reported that small doses of adrenaline give rise to subjective feelings of emotion that resemble normal reactions to real life situations. Subjective feelings evoked by noradrenaline were similar but less intense than those shown following administration of adrenaline. A mixture of the two catecholamines gave results that were equal to an equivalent volume of adrenaline. If measurements of human performance were used when Frankenhaeuser found that there was a positive correlation with the dose level of adrenaline being used. With high concentrations of adrenaline, performance was sometimes affected by hand tremors but these could be overcome by the subject increasing his concentration on the task being performed.

These findings are supported by clinical evidence from patients suffering from a disease called pheochromocytoma in which a benign tumour in the adrenal medulla results in progressively increasing amounts of adrenaline and noradrenaline being released into the circulation. Patients suffering from this disease have episodes of intense sympathetic nervous system activity that are accompanied by anxiety. As the tumour enlarges, the anxiety states become more frequent and are said to become triggered by an increased activity of the sympathetic nervous system (Steinwald *et al.* 1969; Winkler and Smith, 1972).

One major source of confusion in the literature is that small doses of adrenaline produce activity of the sympathetic nervous system which paradoxically decrease as the dose level is increased. It is thought that the small levels induce peripheral effects, while larger doses produce suppression of behaviour via effects on the central nervous system. Gellhorn (1965) has suggested that if the level of sympathetic activity increases beyond a certain level, it flows over and triggers the parasympathetic system causing a trophotropic rebound effect. Gellhorn argues that this 'rebound reaction' is due to the initial brain activity in the posterior area of the hypothalamus, with its predominantly sympathetic effects, spilling over into the anterior areas of the hypothalamus, with its predominantly parasympathetic effects. Breggen (1964) has suggested that larger doses of adrenaline act as a brake preventing the spiralling effects of anxiety. Clearly, such a feedback mechanism would have advantages in damping-down excessive and possibly damaging sympathetic reactions. The point at which this takes place is not too clear; it may take place at the level of the hypothalamus as suggested by Gellhorn and Breggen, or in a centre located in the reticular activating system as suggested by Dell *et al.* (1954). Alternatively, both areas may interact to limit sympathetic overactivity.

CATECHOLAMINES AND HUMAN PERFORMANCE

Frankenhaeuser (1971) has lucidly summarized a series of studies carried out in the Psychological Laboratories of the University of Stockholm, measuring the excretion rate of catecholamines. Within certain limits, it was found that plotting level of adrenaline against performance, gave a dose response curve showing small increases in efficiency. Also, catecholamine levels during moderate activity rose to twice normal resting levels while moderate stress produced increases of up to three to five times normal resting levels. Mental activity produced increases in adrenaline but not noradrenaline. Physical work produced increases in both catecholamines. The effect of visual and auditory stimulation upon subjects was studied in an experiment using two levels of stimulation. The first involved the subject in a 3-hour vigilance task using visual cues. The second task involved the subject simultaneously attending to a visual and an auditory task. Levels of both catecholamines were found to be significantly increased for the twofold task. In view of earlier points made about the effects produced by uncertainty in monkeys, it is interesting to note that Frankenhaeuser reports that uncertainty produced large increases in adrenaline excretion.

In another series of studies, dividing subjects into low and high adrenaline output groups, it was shown that catecholamine output matched behavioural efficiency. Using the Stroop colour-word test that requires close concentration, it was found that a group of subjects having high adrenaline output performed consistently better than a group showing low adrenaline output. Similarly in a choice reaction time experiment, subjects showing high adrenaline output recorded faster reaction times compared to subjects with low adrenaline output. Assays made of the excreted adrenaline levels in habitual 'morning' and 'evening' workers showed that the peaks of the subject's adrenaline output matched his periods of optimal performance. Thus it would appear that catecholamines' output matches performance efficiency and level of alertness. This finding was extended by an experiment in which it was found that night-time resting levels of adrenaline showed a significant positive correlation with intellectual level of young children, as measured by conventional intelligence tests.

Akerstedt and Froberg (1975) have measured catecholamine levels of individuals over extended periods and, using harmonic and rhythm statistical analysis, have revealed a sinusoidal circadian rhythm with maximum values occurring during the day and minimum periods occurring during the night. Peak periods during the day were variable but were usually found between 1200-1500 hours. In terms of adrenaline output the authors felt that it was meaningful to make a distinction between morning and evening alert individuals. Analysis of the circadian rhythms of adrenaline in both two- and three-shift workers showed that most of the original non-shift work pattern, i.e. peaking during the day, was retained even when these workers were on the night-shift. Extended periods of nightwork produced a breaking up of the normal nonshift work pattern with permanent nightworkers showing an inverted function with peak adrenaline output occurring during the night.

The studies carried out in Stockholm implicate adrenaline as an important hormone for coping with a variety of social and psychological stressors and also show that adrenaline excretion is related to cognitive and behavioural performance. However, large individual differences in catecholamine output exist and, as mentioned earlier, these may, in part, relate to previous conditioning and learning.

In general, the correlations for noradrenaline and behaviour are said by Frankenhaeuser to be similar to those of adrenaline, but in the case of noradrenaline they are less consistent and the correlations are smaller. It is possible that mental alertness and activity are related and it would be of interest to tease out these factors and the following studies go someway towards this aim.

3-METHOXY-4-HYDROXPHENYLGLYCOL

The clinical condition mentioned earlier called pheochromocytoma enabled various metabolites or breakdown products of the catecholamines to be determined. Maas and Landis (1965, 1967) reported that 25-30 per cent of the metabolite 3-methoxy-4-hydroxyphenylglycol (MHPG) in dogs' urine represented noradrenaline metabolized in the brain. This initial discovery was followed up by Gitlow et al. (1971) who reported that 20-35 per cent of the MHPG metabolite excreted in the urine of humans arose from cerebral metabolism. Following these discoveries a number of studies have been reported in which the metabolite was assayed in patients suffering from manic depression (Maas et al., 1971; Bond et al., 1972; Fawcett et al., 1972; Maas et al., 1972; Jones et al., 1973). These authors showed that levels of MHPG secreted rose during the manic phases of the manic-depressive cycle and fell during the depressive phases of the cycle. The suggestion made by these authors was that the changing levels reflected the level of brain activity. However, it could be claimed, as with the studies of free catecholamines, that the levels of MHPG reflected alterations of muscular activity during the different phases of the cycle rather than brain activity. Goode et al. (1973) tested this idea by using subjects over a period of several days comparing a rest period with a 2-hour period of isometric and isotonic exercises. These authors found no increases in MHPG levels following the exercise periods and concluded that their results supported the suggestion that the results obtained from manic-depressive patients were not likely to be due to differences in muscular activity. In addition, Maas et al. (1971) have argued that the changes in MHPG levels were not a by-product of concurrent biochemical changes taking place in the brain.

The studies reported above indicate a relationship between noradrenaline metabolism in the brain and cognition. Rubin et al. (1970) reported a study that investigated urinary MHPG levels in naval pilots and their navigators during three training exercises related to landing aircraft on the flight deck of a warship. The phases were: (1) simulated flight deck landings at the shore base; (2) day-time landings on a carrier ship; (3) night-time flight deck landings on the carrier ship. Night-time landings on a moving carrier are said to be the most difficult and complicated task demanded from the naval pilots. They call for a number of complex perceptual-motor skills and precise spatial orientation. The levels of MHPG found in the urine of the pilots and navigators were shown to be commensurate with the difficulty of the landings. In all cases the pilots were said to show higher levels than their navigators. Clearly in such a dangerous situation there is a strong stress component, for the dangers are very apparent and real. In considering this point Rubin and his collaborators point out that, in an earlier study which examined the levels of cortisol in the blood, it was the pilots who showed increases on flying days and not the navigators. The authors finally concluded that the MHPG changes appeared to be related to the intensity of concentration, attention, and alertness required to perform the task and further-more, these factors did not diminish with experience.

Frankenhaeuser *et al.* (1976) have reported a study in which 4-hydroxy-3-methoxyphenylethylene glycol (MOPEG) excretion levels were analysed in male and female students. Urinary excretion of cortisol, adrenaline, noradrenaline, and MOPEG were found to increase in both sexes. In the male group the levels of adrenaline and MOPEG increased significantly. Both groups performed equally well in the examination; however, self-report questionnaires revealed that males felt confident and successful whereas the females reported discomfort and lack of confidence.

The discovery of the metabolite MHPG appears to hold a promise for future research and further detailed examination of the relationship between cognition and the metabolism of noradrenaline is required. Although at present, the assay techniques pose research problems that need to be overcome.

THE CATECHOLAMINE HYPOTHESES OF MOOD

As indicated in Chapter 3, dopamine and noradrenaline play important roles in brain function; indeed, they have been argued to have crucial roles in regulating mood and feeling states. This mood function is called the catechol-amine hypothesis of affective states (Kety, 1966; Schildkraut, 1965, 1973; Schildkraut and Kety, 1967). Basically, the hypothesis states that the level of catecholamines in the brain determines the mood of the person. High levels are correlated with euphoria and low levels with depression. Pharmacological treatment of manic patients consists of drugs that will block or depress the biosynthesis of noradrenaline; for example, in Figure 4.4 the drug reserpine is shown as having a negative effect on the levels of intraneuronal noradrenaline. Conversely, pharmacological treatment of psychological depression consists of treatment with a drug that allows the build-up of noradrenaline within the neurone or slows down its destruction following its release. In Figure 4.4 monoamine oxidase (MAO) is shown as an enzyme that breaks down intracellular noradrenaline. The drug iproniazid is used to produce inhibition of MAO allowing levels of intracellular noradrenaline to increase. In European countries control of brain catecholamine levels has been mainly carried out via modifications of the action of MAO. Friedhoff (1975) and Mendels (1975) have both indicated that the simple catecholamine hypothesis of affective state is no longer tenable in its original form. As Mendels has pointed out mania and depression are not opposite states, both complaints involve a number of common biological changes and manic patients may show signs and symptoms of depression while in a manic phase of their illness.

From animal studies using intracranial self-stimulation and similar techniques Stein and his coworkers (Stein *et al.*, 1973) have proposed a reciprocal mechanism in the brain that involves neurochemicals rather than brain areas. The neurochemicals involved are the biogenic amines noradrenaline and serotonin. Their proposal is that the 'reward' systems of the brain, centred around the medial forebrain bundle (see Figure 4.4), are mediated by noradrenaline while the 'punishment' systems, involving the periventricular areas and pathways (see Figure 4.4), are mediated by serotonin. The proposal is congruent with the view that elation arises from increased activity of noradrenaline in the brain. Originally it was proposed that acetylcholine was the transmitter substance mediating the periventricular system (Stein, 1969) but the switch to serotonin (Wise *et al.*, 1973) has enabled them to explain a number of contradictions in both animals and human studies involving drugs. They suggest that clinically depressed patients have low levels of noradrenaline but in their suggested reciprocal system this means there will be a relatively larger amount of serotonin present in the brain, involving a corresponding increase in the activity of the 'punishment' systems. This is their reason why it is often found that depressed patients show high levels of anxiety. If their theory is shown to be correct, then drug treatment of depressed patients would involve selective alteration of both of these important biogenic amines rather than just one.

Lower brain areas contain large amounts of noradrenaline that may relate to cortical activity and also form feedback loops with the peripheral nervous system. Weil-Malherbe (1960) has demonstrated that radioactivity labelled noradrenaline can slowly penetrate the blood-brain barrier in the area of the hypothalamus and this would also tie in with the concept of a sympathetic/parasympathetic balance mediated by the hypothalamus.

An interesting recent development is the discovery by Porter *et al.* (1965) that the compound 6-hydroxy-dopamine (6-OHDA) results in prolonged depletion of noradrenaline from sympathetic neurones and innervated organs. Angeletti (1971) has claimed that injections of this compound into neonatal rats produces almost complete hypotrophy of the sympathetic nervous system. Evetts *et al.* (1970) have indicated that 6-OHDA results in prolonged depression of noradrenaline levels in brain neurones and Laverty and Arnott (1970) showed that a central intraventricular injection of 6-OHDA into adult rats resulted in reduced reactions in a learned avoidance task.

FUTURE TRENDS IN PSYCHOPHARMACOLOGY

What are the likely trends in this area? Carruthers and Taggart (1975) have suggested that one of the most important roles of noradreanline is its stimulant action on the 'pleasure' centres of the brain. They claim that physical exercise is the natural way of increasing noradrenaline levels in the body and when exercise is taken the body also utilizes the additional levels of glucose and blood lipids that are released at the same time as noradrenaline. Industrial and urban life have tended to result in reduced opportunity for many people to engage in physical exercise and as a consequence, increasing numbers have tended to use drugs such as the nicotine in cigarettes or the caffeine in tea and coffee for these drugs exert a stimulating action on noradrenaline. Unfortunately, these drugs do not result in the utilization of the glucose and blood lipids that are released with noradrenaline. Carruthers and Taggart feel that one of the most important aspects in the man-and-his-environment interaction will be the need to replace these commonly used and abused drugs with exercise and relaxation techniques.

Up to the present time drugs have been rather gross instruments and few are free from damaging side effects that in some cases have been reported as causing more problems than the original complaint they were designed to cure. However, progress is being made and we can increasingly expect to find drugs that are more selective in their actions. An important area, concerning the medical aspects of the autonomic nervous system, is the further development of selective alpha- and beta-blockers to control high blood pressure.

Perhaps the future of pharmacology for the psychologist who is interested in recording from animals and humans engaged in activity lies in the development of electrochemical techniques (Adams, 1976). Electrochemical techniques of voltammetry and chronamperometry should enable almost instantaneous determinations about the release of transmitter substances in fairly precise locations in the nervous system. From such studies we will be able to build up a picture of the chemical pathways in the brain and peripheral nervous system that will serve to complement our existing knowledge about the electrical pathways in the brain and peripheral nervous system.

REFERENCES

Adams, R.N. (1976). 'Probing brain chemistry with electroanalytic techniques', *Anal. Chem.*, **48**, 1128-38.

Akerstedt, T. and Froberg, J.E. (1975). 'Circadian rhythms: catecholamine excretion, performance and alertness', Paper given at the Workshop on Catecholamines and Behaviour. (Convenor) M. Frankenhaeuser, Stockholm, May 1975.

Alquist, R.P. (1948). 'A study of the adrenergic receptors', *Amer. J. Physiol.*, **153**, 586-600.

Altman, L.L., Pratt, D. and Cotton, J.M. (1943). 'Cardiovascular response to acetyl-beta-methylcholine (Mecholyl) in mental disoders', *J. Nervous and Mental Diseases*, **97**, 296-309.

Angeletti, P.U. (1971) 'Chemical sympathectomy in newborn animals', *Neuropharmacol.*, **10**, 55-9.

Ax, A.F. (1953). 'The physiological differentiation between fear and anger in humans', *Psychosomatic Medicine*, **15**, 433-42.

Ax, A.F. (1964). 'Goals and methods in psychophysiology', *Psychophysiol.*, **1**, 5-25.

Becker, R.O., Bachman, C.H. and Friedman, H. (1962). 'The direct current control system, a link between environment and organism. *New York State J. Medicine, 15* April, 1169-76.

Benjamin, L.S. (1963). 'Statistical treatment of the Law of Initial Value in autonomic research: a review and recommendation', *Psychosomatic Medicine*, **25**, 556-66.

Benjamin, L.S. (1967). 'Facts and artefacts in using analysis of covariance to "undo" the Law of Initial Value', *Psychophysiol.*, **4**, 187-202.

Binik, Y.M., Theriault, G. and Shustack, B. (1977). 'Sudden death in the laboratory rat: cardiac function, sensory and experiential factors in swimming deaths', *Psychosomatic Medicine*, **34**, 82-92.

Bond, P.A., Jenner, F.A. and Simpson, G.A. (1972). 'Daily variations of the urine content of 3-methoxy-4-hydroxyphenylglycol in two manic-depressive patients', *Psychological Medicine*, **2**, 81-85.

Brady, J.V. (1967). 'Emotion and sensitivity of psychoendocrine systems', in D.C. Glass (Ed.) *Neurophysiology and Emotion.* Rockefeller University Press.

Breggen, P.R. (1964). 'The psychophysiology of anxiety: with a review of the literature concerning adrenaline', *J. Nervous and Mental Diseases*, **139**, 558-68.

Broadbent, D. (1977). 'Levels, hierarchies, and the locus of control', *Quart. J. Exp. Psychol.*, **29**, 181-201.

Brown, C.C. (1966). 'Psychophysiology at an interface. *Psychophysiol.*, **3**, 1-7.

Brown, C.C. (1967). *Methods of Psychophysiology.* Williams and Wilkins, Baltimore.

Burnstock, G. (1969). 'Evolution of the autonomic innervation of visceral and cardio-vascular system in vertebrates', *Pharmacol. Rev.*, **21**, 247-324.

Campos, J.J. and Johnson, H.J. (1967). 'Affect, verbalization, and directional fractionation of autonomic responses', *Psychophysiol.*, **3**, 285-290.

Carruthers, M. and Taggart, P. (1975). 'Man in noradrenaline secreting states'. Paper given at the Workshop on Catecholamines and Behaviour. (Convenor) M. Frankenhaeuser. Stockholm, May 1975.

Darrow, C.W. (1943). 'Physiological and clinical tests of autonomic function and autonomic balance', *Physiol. Rev.*, **23**, 1-36.

Darrow, C.W., Jost, H., Solomon, A.P. and Mergener, J.C. (1942). Autonomic indications of excitatory and homeostatic effects on the electroencephalogram. *J. Psychol.*, **14**, 115-30.

Dell, P., Bonvallet, M. and Hugelin, A. (1954). 'Tonus sympathique adrenaline et controle reticulaire de la motricite spinale', *EEG Clin. Neurophsyiol.*, **6**, 599-618.

Duffy, E. (1934). 'Emotion: an example of the need for reorientation in psychology', *Psychol. Rev.*, **41**, 184-98.

Duffy, E. (1972). 'Activation', in N.S. Greenfield and R.A. Sternbach, (Eds.) *Handbook of Psychophysiology.* Holt, Rhinehart.

Elliott, R. (1964). 'Physiological activity and performance; a comparison of kindergarten children with young adults', *Psychol. Monogr.* **78**, No. 10.

Eppinger, J. and Hess, L. (1915). 'Die vagotonie', *Mental and Nervous Disease Monogr, No. 20.*

Euler von, U.S. (1964). 'Quantification of stress by catecholamine analysis', *J. Clin. Pharmacol. Therapy*, **5**, 398-404.

Evetts, K.D., Uretsky, N.J., Iversen, L.L. and Iversen, S.D. (1970). 'Central nervous system effects of 6-hydroxydopamine', *Nature*, **225**, 961-2.

Fawcett, J., Maas, J.W. and Dekirmenjian, H. (1972). 'Depression and MHPG excretion', *Arch. Gen. Psychiat.*, **26**, 246-251.

Feinberg, I. (1958). 'Current status of the Funkenstein test', *Arch. Neurol. Psychiat.*, **80**, 488-501.

Frankenhaeuser, M. (1971). 'Behaviour and circulating catecholamines', *Brain Res.*, **31**, 241-62.

Frankenhaeuser, M., Von Wright, M.R., Collins, A., Von Wright, J., Sedvall, G. and Swahn, C-G. (1976). 'Sex differences in psychoneuroendocrine reactions to examination stress', *Report No. 489, Department of Psychology, University of Stockholm.*

Friedhoff, A.J. (ed.) (1975). *Catecholamines and Behaviour 2. Neuropharmacology,* Plenum Press, New York.

Funkenstein, D.H. (1956). 'Norepinephrine-like and epinephrine-like substances in relation to human behaviour', *J. Nervous and Mental Disease,* **124**, 58-66.

Gellhorn, E. (1965). 'The neurophysiological basis of anxiety: a hypothesis', *Perspectives in Biology and Medicine,* **8**, 488-515.

Gellhorn, E. and Loofbourrow, G.N. (1963). *Emotions and Emotional Disorders: a Neurophysiological Study.* Harper and Row, New York.

Gitlow, S.E., Mendlowitz, M., Bertani, L.M., Wilk, S. and Wilk, E.K. (1971). 'Human norepinephrine metabolism. Its evaluation by administration of tritiated norepinephrine', *J. Clin. Invest.,* **50**, 859-65.

Glickman, S.E. and Schiff, B.B. (1967). 'A biological theory of reinforcement', *Psychol. Rev.,* **74**, 81-100.

Goldstein, A., Aronow, L. and Kalman, S.M. (1974) *Principles and Drug Action: The Basis of Pharmacology* (2nd ed.). John Wiley and Sons, New York.

Goodall, McC. (1951). 'Studies of adrenaline and noradrenaline in mammalian heart and suprarenals', *Acta Physiological Scandinavia,* **24**, Supplement 85.

Goodall, McC (1962). 'Sympathoadrenal response to gravitational stress', *J. Clin. Invest.,* **41**, 197-202.

Goode, D.J., Dekirmenjian, H., Meltzer, H.Y. and Maas, J.W. (1973). 'Relation of exercise to MHPG excretion in normal subjects', *Arch. Gen. Psychiat.,* **29**, 391-6.

Goodman, L.S. and Gilman, A. (1970). *The Pharmacological Basis of Therapeutics for Physicians and Medical Students* (4th ed.). Macmillan, New York.

Goth, A. (1974). *Medical Pharmacology,* C.V. Mosby, St. Louis.

Greenfield, N.S. and Sternbach, R.A. (Eds.) (1972). *Handbook of Psychophysiology.* Holt, Rinehart and Winston, London.

Grings, W.H. (1977). 'Orientation, conditioning and learning', *Psychophysiol.,* **14**, 343-50.

Grossman, S.P. (1973). *Essentials of Physiological Psychology,* John Wiley, New York.

Gunderson, E.K. (1953). *Autonomic Balance in Schizophrenia.* Unpublished Ph.D. thesis, University of California.

Hahn, W.W. (1973). 'Attention and heart-rate: a critical appraisal of the hypothesis of Lacey and Lacey', *Psychol. Bull.,* **79**, 59-70.

Hersch, R.G., Antrobus, J.S., Arkin, A.M. and Singer, J.L. (1970). 'Dreaming as a function of sympathetic arousal', *Psychophysiol.,* **7**, 329-30.

Hines, E.A. and Brown, G.E. (1933). 'Standard test for measuring variability of blood pressure. Its significance as an index of prehypertensive state', *Ann. Int. Med.,* **7**, 209-217.

Hohmann, G.W. (1966). 'Some effects of spinal cord lesions on experienced emotional feelings', *Psychophysiol.,* **3**, 143-56.

Igersheimer, W.W. (1953). 'Cold pressor test in functional psychiatric syndromes', *Arch. Neurol. Psychiat.,* **70**, 794-801.

Iversen, L.L. (1967). *The Uptake and Storage of Noradrenaline in Sympathetic Nerves.* Cambridge University Press, London.

Iversen, L.L. (1975). 'Dopamine receptors in the brain', *Science,* **188**, 1084-9.

Jenkinson, D.H. (1973). 'Classification and properties of peripheral adrenergic receptors', in L.L. Iverson (Ed.) *Catecholamines. Brit. Med. Bull.,* **29**,.

Johnson, L.C. (1970). 'A psychophysiology for all states', *Psychophysiol.* **6**, 501-16.

Johnson, R.H. and Spalding, J.M.K. (1974). *Disorders of the Autonomic Nervous System.* Blackwell, Oxford.

Jones, F.D., Maas, J.W., Dekirmenjian, H. and Fawcett, J. (1973). 'Urinary catecholamine metabolites during behavioural changes in a patient with manic depressive cycles', *Science,* **179**, 300-2.

Kety, S.S. (1966). 'Catecholamines in neuropsychiatric states', *Pharmacol. Rev.*, **18**, 787-98.

Lacey, J.I. (1950). 'Individual differences in somatic response patterns', *J. comp. physiol. Psychol.*, **43**, 338-50.

Lacey, J.I. (1956). 'The evaluation of autonomic responses: towards a general solution', *Ann. New York Academy of Sciences*, **67**, 123-64.

Lacey, J.I. (1967). 'Somatic response patterning and stress: some revisions of activation theory', in M.H. Appley and R. Trumball (Eds.) *Psychological Stress*. Appleton-Century-Crofts, New York.

Lacey, J.I. and Lacey, B.C. (1962). 'The law of initial values in longitudinal studies of autonomic constitution. Reproducibility of autonomic nervous system response patterns over a four year interval'. *Ann. New York Academy of Science*, **98**, 1257-1290.

Lacey, J.I. and Lacey, B.C. (1970). 'Some autonomic-central nervous system inter-relationships', in P. Black (Ed.) *Physiological Correlates of Emotion*. Academic Press, New York.

Lader, M. (1975) *The Psychophysiology of Mental Illness*. Routledge and Kegan Paul, London.

Laverty, R. and Arnott, P.J. (1970). 'Recovery of avoidance behaviour in rats following intraventricular injection of 6-hydroxydopamine', *Proc. of the University of Otago Medical School*, **48**, 18-20.

Levi, L. (1965). 'The urinary output of adrenaline and noradrenaline during pleasant and unpleasant emotional states. A preliminary study', *Psychosomatic Medicine*, **27**, 80-85.

Lindsley, D.B. (1951). Emotion, in S.S. Stevens (Ed.) *Handbook of Experimental Psychology*. J. Wiley and Sons, New York.

Lindsley, D.B. (1970). The role of non-specific reticulo-thalamo-cortical systems in emotion', in P. Black (Ed.) *Physiological Correlates of Emotion*. Academic Press, New York.

Lindsley, D.B., Bowden, J. and Magoun, H.W. (1949). 'Effect upon the EEG of acute injury to the brain stem activating system', *EEG Clin. Neurophysiol.* **1**, 475-86.

Lindsley, D.B. Schrines, L.H., Knowles, W.B. and Magoun, H.W. (1950). 'Behavioural and EEG change following chronic brain stem lesions in the cat', *EEG Clin. Neurophysiol.*, **2**, 483-98.

Longo, V.G. (1972). *Neuropharmacology and Behaviour*. Freeman, San Francisco.

Lovallo, W. (1975) 'The cold pressor test and autonomic function: a review and integration', *Psychophysiol.*, **12**, 268-83.

Lynn, R. (1966) *Attention, Arousal, and the Orientation Reaction*, Pergamon Press.

Maas, J.W., Dekirmenjian, H. and Fawcett, J. (1971). 'Catecholamine metabolism in depression and stress', *Nature*, **230**, 330-1.

Maas, J.W. Fawcett, J.A. and Dekirmenjian, H. (1972). 'Catecholamine metabolism, depressive illness and drug response', *Arch. Gen. Psychiat.*, **26**, 252-62.

Maas, J.W. and Landis, D.H. (1965). 'Brain norepinephrine: a behavioural and kinetic study', *Psychosomatic Medicine*, **27**, 399-407.

Maas, J.W. and Landis, D.H. (1967). '*In vivo* studies of the rates of disappearance in urine of metabolites of brain norepinepharine', *Federation Proc.*, **26**, 463.

McGuigan, F.J. and Schoonover, R.A. (1973). *The Psychophysiology of Thinking: Studies of Covert Processes*. Academic Press, New York.

McKilligott, J.WV. (1959). *Autonomic Functions and Affective States in Spinal Cord Injury*. Unpublished Ph.D. thesis. University of California.

Malmo, R.B. (1965). Finger-sweat prints in the differentiation of low and high incentive. *Psychophysiol.*, **1**, 231-40.

Malmo, R.B. (1972). 'Overview', in N.S. Greenfield and R.A. Sternbach (Eds.) *Handbook of Psychophysiology*. Holt, Rinehart, and Winston, New York.

109

Mandler, G. (1967). 'The conditions for emotional behaviour', in D.C. Glass (Ed.) *Neurophysiology and Behaviour*. Rockefeller University.

Marañon, G. (1924). 'Contribution a l'etude de l'action emotive de l'adrenline', *Revue Francaise d'Endcrinologie*, **2**, 301-25.

Mendels, J. (1975). *The Psychobiology of Depression*. Spectrum Press, New York.

Monnier, M. (1968). *Functions of the Nervous System Volume 1 Autonomic Functions*. Elsevier Publishing Co., Amsterdam.

Moruzzi, G. and Magoun, H.W. (1949). 'Brainstem reticular formation and activation of the EEG, *EEG Clin. Neurophysiol.*, **1**, 455-73.

Obrist, P.A. (1976). 'The cardiovascular-behavioural interaction as it appears today', *Psychophysiol.*, **13**, 95-107.

Obrist, P.A., Black, A.H., Brener, J. and DiCara, L.V. (Eds.) (1974). *Cardiovascular Psychophysiology*. Aldine, Chicago.

Orme-Johnson, D.W. (1973). 'Autonomic stability and transcendental meditation', *Psychosomatic Medicine*, **35**, 341-9.

Plutchik, R. and Ax, A.F. (1967). 'A critique of determinants of an emotional state by Schachter and Singer (1962)', *Psychophysiol.* **4**, 79-82.

Porter, C.C., Totaro, J.A. and Burcin, A. (1965). 'The relationship between radioactivity and noradrenaline concentration in the brain and heart of mice following administration of labelled methyl dopa or 6-hydroxydopamine', *J. Pharmacol. Exp. Therapeutics*, **150**, 17-22.

Richter, C.P. (1957). 'On the phenomenon of sudden death in animals and man', *Psychosomatic Medicine*, **19**, 191-8.

Roessler, R. and Greenfield, N.S. (1962) *Physiological Correlates of Psychological Disorder*. University of Wisconsin Press.

Routenberg, A. (1968). 'The two-arousal hypothesis: reticular formation and limbic system', *Psychol. Rev.*, **75**, 51-80.

Rubin, R.T., Miller, R.G., Clark, B.R., Poland, R.E. and Ranson, J.C. (1970). 'The stress of aircraft carrier landings II, 3-methoxy-4-hydroxyphenylglycol excretion in naval aviators', *Psychosomatic Medicine*, **32**, 589-97.

Schachter, S. and Singer, J.E. (1962) 'Cognitive, social, and physiological determinants of emotional state', *Psychol. Rev.*, **69**, 379-99.

Schildkraut, J.J. (1965). 'The catecholamine hypothesis of affective disorders: a review of supporting evidence', *Amer. J. Psychiat.*, **122**, 509-22.

Schildkraut, J.J. (1973). 'Neuropharmacology of the affective disorders', *Ann. Rev. Pharmacol.*, **13**, 427-54.

Schildkraut, J.J. and Kety, S.S. (1967). 'Biogenic amines and emotion', *Science*, **7**, 21-30.

Sherry, L.U. (1959). *Some Effects of Chlorpromazine on the Physiological and Psychological Functioning of a Group of Schizophrenics*. Unpublished Ph.D. thesis. University of California.

Sjarne, L., Hedquist, P. and Bygoleman, S. (1969). 'Neurotransmitter quantum released from sympathetic nerves in the cat's skeletal muscle', *Life Science*, **8**, 198-96.

Sokolov, E.N. (1960). 'Neuronal models and the orienting reflex', in M.A.B. Brazier (Ed.), *The Central Nervous System and Behaviour*. Macy Foundation, New York.

Sokolov, E.N. (1963a). *Perception and the Conditioned Reflex*. Pergamon, London.

Sokolov, E.N. (1963b). 'Higher nervous functions: the orienting reflex', *Ann. Rev. Physiol.*, **25**, 545-80.

Stein, L. (1969). 'Chemistry of reward and punishment', in D.H. Efron (Ed.) *Psychopharmacology: a Review*. U.S. Government Printing Office, Washington, D.C.

Stein, L., Wise, C.D. and Berger, B. (1973). 'Antianxiety action of benzodiazepines: decrease in activity of serotonin neurons in the punishment system', in S. Garatinni, E. Mussini and Randall, L.O. (Eds.) *The Benzodiazepines*. Ravens Press, New York.

Steinwald, O.P., Doolas, A. and Southwick, H.W. (1969). 'Pheochromocytoma', *Surgical Clinics of North America*, **49**, 87-98.

Stern, J.A. (1972). 'Physiological response measures during classical conditioning', in N.S. Greenfield and R.A. Sternbach, (Eds.) *Handbook of Psychophysiology*. Holt, Rinehart, and Winston, New York.

Sternbach, R.A. (1960). 'The independent indices of activation', *EEG Clin. Neurophysiol.*, **12**, 609-11.

Sternbach, R.A. (1966). *Principles of Psychophysiology*. Academic Press, London.

Stroebel, C.F. (1972). 'Psychophysiological pharmacology', in N.S. Greenfield and R.A. Sternbach (Eds.) *Handbook of Psychophysiology*, Holt, Rinehart, and Winston, New York.

Surwillo, W.W. and Arenburg, D.L. (1965). 'On the Law of Initial Values and measures of change', *Psychophysiol.* **1**, 368-70.

Tart, C.T. (Ed.) (1969). *Altered States of Consciousness*. Wiley, New York.

Uno, T. and Grings, W.W. (1965). 'Autonomic components of orientating behaviour', *Psychophysiol.*, **1**, 311-21.

Vander, A.J., Sherman, J.H. and Luciano, D.S. (1970). *Human Physiology: The Mechanisms of Bodily Function*. McGraw Hill, London.

Venables, P.H. and Christie, M.J. (Eds.) (1975). *Research in Psychophysiology*, J. Wiley and Sons, London.

Venables, P.H. and Martin, I. (Eds.) (1967). *Manual of Psychophysiological Methods*. Wiley, New York.

Wallace, R.K., Benson, H. and Wilson, A.F. (1971). 'A wakeful hypometabolic physiologic state', *Amer. J. Physiol.*, **221**, 795-9.

Warburton, D.M. (1975). *Brain, Behaviour, and Drugs: An Introduction to the Neurochemistry of Behaviour*. J. Wiley and Sons, London.

Webb, R.A. and Obrist, P.P. (1970). 'The physiological concomitants of reaction time performance as a function of preparatory interval and preparatory interval series', *Psychophysiol.*, **6**, 389-403.

Weil-Malherbe, H. (1960). 'The passage of catecholamines through the blood-brain barrier', in J.R. Vane (Ed.) *Adrenergic Mechanisms*. Ciba Foundation Symposium. Churchill, London.

Wenger, M.A. (1941). 'The measurement of individual differences in autonomic balance', *Psychosomatic Medicine*, **3**, 427-34.

Wenger, M.A. (1948). 'Studies of autonomic balance in U.S.A.F. personnel', *Comparative Psychol. Monog.*, **19**, No. 4.

Wenger, M.A. (1966). 'Studies of autonomic balance: a summary', *Psychophysiol.*, **2**, 173-86.

Wenger, M.A., Jones, F.N. and Jones M.H. (1956). *Physiological Psychology*. Holt, New York.

Wise, C.D., Berger, B.D. and Stein, L. (1973). Evidence of noradrenergic reward receptors and serotonergic punishment receptors in the rat brain. *Biol. Psychiat.*, **6**, 3-21.

Wilder, J. (1957). 'The Law of Initial Values in neurology and psychiatry. Facts and problems', *J. Nervous and Mental Disease*, **125**, 73-86.

Wilder, J. (1962). 'Basimetric approach (Law of Initial Value) to biological rhythms', *Ann. New York Academy Science*, **98**, 1211-20.

Winkler, H. and Smith, A.D. (1972). 'Pheochromocytoma and other catecholamine producing tumours', in H. Blaschko and E. Muscholl (Eds.) *Catecholamines*. Springer-Verlag, Berlin.

Zimny, G.H. and Kienstra, R.A. (1967). 'Orienting and defensive responses to electric shock', *Psychophysiol.*, **3**, 351-62.

Zimny, G.H. and Miller, F.L. (1966). 'Orienting and adaptive cardiovascular responses to heat and cold', *Psychophysiol*, **3**, 81-92.

Zimny, G.H. and Schwabe, L.W. (1966). 'Stimulus change and habituation of the orienting response', *Psychophysiol.*, **2**, 103-15.

CHAPTER 5

Biofeedback and Emotion

In recent years attempts have been made to condition internal physiological processes that are generally held to be outside of the influence of conscious control, many of these attempts involve the autonomic nervous system. Described as biofeedback techniques these attempts have not only been applied to autonomic functions but also to activities of the brain and the level of muscular tension; however, these brain and somatic experiments will only be considered in so far as they relate to autonomic functions. References to biofeedback studies can be found in the Aldine overviews on biofeedback and self-control that are published annually (Barber *et al.*, 1977).

Even allowing for such specialists as yogi practitioners (Wenger *et al.* 1961), it has been common knowledge for many years that certain individuals were able to exert considerable control over their autonomic nervous system. William Falconer wrote a book in 1796 with the title *The Influence Of The Passions Upon Disorders Of The Body*, which dealt with relationships between diseases and emotion. In 1800, Cruttwell's of Bath, England, published a monograph by John Haygarth, *Of the Imagination, as a Cause and as a Curse of the Disorders of the Body*: exemplified by fictions, tractors, and epidemical convulsions in which the author argued the case for imagination as a cause and cure for physical illness. This short tract arose out a paper read to the Literary and Philosophical Society of Bath. Later in 1872, Daniel Tuke wrote a book in which the title, *Illustrations of the Influence of the Mind upon the Body in Health and Disease Designed to Elucidate the Action of the Imagination*, serves as a brief summary of the contents. Tuke discussed control over the voluntary and involuntary functions of the body. During the early part of this century, studies were reported of individuals who were able to exert some level of control over their heart-rates (Favill and White, 1917; Taylor, 1922). Despite these early studies, the idea became firmly accepted that the functions of the autonomic nervous system were beyond conscious control. As pointed out at the beginning of this book, the use of the word autonomic by Langley was perhaps one important factor but it was by no means the only one; the influential physiologist Sherrington held the autonomic nervous system to be beyond conscious control.

The basic physiological argument divided the nervous system into two discrete parts. The brain or central nervous system was concerned with the voluntary functions of the body which were under conscious control. The autonomic nervous system was concerned with the involuntary functions of the body that could not be controlled consciously.

The story is long and complicated but as far as psychology is concerned, this physiological dichotomy seems to have been finally accepted when it appeared to be shown that the two main techniques used in the laboratory to study the learning processes reflected a basic underlying physiological division. However, before we discuss this we need to digress a little into learning theory. Operant techniques pioneered in the Western World by workers such as Thorndike and Skinner were held to reflect learning processes involving the cortical areas of the brain. Operant learning situations usually require animals to behave actively and investigators using these methods invented apparatus such as puzzle boxes, mazes, or operant chambers to analyse the learning processes. The dependent variable or measure of learning in operant learning situations is typically activity measured as response rate or speed. Russian methods characteristically placed the animal into a situation where it was a relatively passive recipient of stimuli being delivered by the experimenter. The classical paradigm of this method was Pavlov's use of the salivary response in dogs to study the learning process. In Pavlov's experiments a dog was restrained in a harness and the dependent variable was rate or volume or salivation produced by a stimulus (the conditioned stimulus) that had previously been linked or paired with food (the unconditioned stimulus). Pavlov's method was held to relate to reflex functions mediated by the autonomic nervous system. For details of specific learning theory techniques and the underlying theoretical issues see Tarpy (1975).

The most complete psychological explanation of what came to be called the two-factor theory of learning appears to have been made by Mowrer (1947) linking together two strange psychological bedfellows, Pavlov and Freud. Mowrer took Pavlov's experimental paradigm linking the conditioned stimulus to an unconditioned stimulus (food, or the avoidance of pain) and added Freud's theoretical idea that anxiety had a subjective and an objective component. Thus, Freud made a fundamental contribution to an area of psychology that prided itself on its strict experimental methodology; moreover, with subtle irony, the main element of his contribution related to a subjective state of anxiety. The reason for Mowrer's theory was the need for learning theorists to explain why it was that, when negative reinforcement (this usually meant electric shock) had been used to condition an animal, removal of the negative reinforcement (Freud's objective anxiety) did not lead to the animal showing an extinction curve (gradual loss of the learning task) found in learning situations where positive reinforcement (food or water) had been used. Indeed, in learning tasks involving avoidance, it is not unusual for animals to continue to respond for many hundreds of trials after the negative reinforcement has been terminated.

Mowrer argued that in this type of learning situation it was usual for the animal to be able to avoid receiving the electric shock by satisfying a criterion imposed by the experimenter. For example, the shock (the unconditioned stimulus) would be signalled by a tone or light (the conditioned stimulus) which lasted for a certain period of time. If the animal made the required response before the conditioned stimulus ended, it avoided the shock for that trial. If it did not avoid the shock by running to the conditioned stimulus it received shock until the response had been made; that is, until the animal escaped from the aversive stimulus. Typically, in this type of learning situation, an animal escapes shock for the first 15-20 trials and then comes to anticipate or avoid the shock by making its response during the warning stimulus. Mowrer's concept was that, initially, when the animal was being shocked, the pain and fear invoked an autonomic reaction (the unconditioned response; Freud's subjective anxiety) that became linked to the tone or light that was being used as the conditioned stimulus (Schoenfeld, 1964). During later trials, when the animal was making avoiding responses and not showing the expected extinction curve, it was because reinforcement was occurring from elimination of the bodily autonomic reactions that were gradually building-up during presentation of the conditioned stimulus. Mowrer's theory contained an explanation for both parts of the nervous system during escape and avoidance learning. During the initial active or escape learning phase there was conscious activity mediated by the brain and, during the later escape trials, there was unconscious activity mediated by the autonomic nervous system.

The above digression should not be thought to be entirely historical because two-factor theory with its emphasis on the role of autonomic function is still used extensively in the clinical and human fields to explain the acquisition of anxiety. For later modifications to two factor theory, interested readers are referred to Mowrer's later book and three excellent reviews (Mowrer, 1960; Rescorla and Solomon, 1967; Herrnstein, 1969; Rescorla and Wagner, 1972).

Currently, it is usual for psychologists working in the area of learning theory to recognize that the two main techniques used to study learning are interrelated. Pavlov himself pointed out that his restrained dogs sometimes used their snouts to knock the tube used to deliver food in order to dislodge trapped particles, i.e., the dogs were making an operant response. In the operant chamber the sight of the lever may be a conditioned stimulus to press the bar etc. However, during the 1930s a number of attempts were made to show that the two methods were theoretically different and quite distinct (Skinner, 1935). Skinner (1938) for example, wrote, 'We may reinforce a man with food whenever "he turns red" we cannot in this way condition him to blush.' Paradoxically in the same article, Skinner points out that 'children seem to learn to cry'. Kimmel (1974) has written an excellent historical overview related to attempts to use operant techniques to condition autonomic systems.

The sum total of these physiological and psychological deliberations, as far as psychology was concerned, was that it became generally accepted that the autonomic nervous system was not under conscious control and as a consequence not amenable to the type of learning experiments carried out by Western psychologists. In short, you could not condition the autonomic nervous system using operant techniques. In its most extreme form it was argued that the autonomic system was a purely outflowing system with no feedback to the brain (Smith, 1954). This view was in keeping with behaviourists' traditional edict, 'it must be external and verifiable'. Despite intermittent discussions, usually supporting the earlier point of view, no serious work was done until Miller (1967, 1969) and his students began a series of experimental studies with the avowed aim of determining if the autonomic nervous system could be conditioned by operant techniques. It was clear that if appropriate techniques could be developed, apart from the theoretical importance, they might also have a tremendous vital role in the treatment of certain psychosomatic illnesses. As will become apparent subsequently, it was this latter aspect that was subjected to a considerable 'oversell' to the detriment of experimental work in this area as a whole (Blanchard and Young, 1973). 'Self-control of cardiac functioning: a promise as yet unfulfilled'. A good example of this overoptimism and unscientific approach is a book by Jonas (1973). More scientifically based but equally sensational are books by Karlins and Andrews (1972) and Brown (1974).

Attempts to use operant techniques to condition functions of the autonomic nervous system fall neatly into animal experiments in which we cannot be sure if cognitive factors are operating and human experiments in which we know that there are strong cognitive factors operating. These two aspects will be discussed separately and in both cases the type of experiments used will be briefly reviewed before considering problems associated with the studies.

THE CONTRIBUTION OF ANIMAL EXPERIMENTS TO BIOFEEDBACK

Biofeedback is basically concerned with self-control over internal physiological events and the particular contribution from animal experiments concerns the strict control procedures that can be introduced in these studies. Also drugs, such as curare, can be used to paralyse the muscular system to prevent subvert muscle action from inducing the particular autonomic change that is being studied.

Initial experiments were reported by Miller and Carmona (1967) who conditioned an operant salivary response in rats using water as the reinforcement. The classical objection to earlier claims for successful conditioning of autonomic functions was that the changes had been induced by subvert muscular activity. This appeared to be the case in this study for bursts of salivation were matched by increases in activity. One way of eliminating or reducing interfering effects of muscular activity is to use a drug that suppresses muscular activity, and it was for this reason curare was used in this study. Unfortunately it was discovered that the drug itself produced copious levels of salivation that obliterated the previous result. Perhaps this would have been enough to discourage most investigators, but Miller, displaying the ingenuity and tenacity that has been the hallmark of his scientific work, set about using other physiological systems and more sophisticated experimental designs. In the next series of studies (Trowill, 1967) an attempt was made to condition heart-race in curarized rats using intracranial self-stimulation (ICSS) as reinforcement. The experiment used curare to paralyse the muscular systems of rats. By altering their heart-rates the curarized rats were able to reinforce themselves with minute electrical currents in the general area of their medial forebrain bundle (see Figure 3.6). The experimental design involved half of the animals being rewarded for increases in heart-rate and half the animals rewarded for decreases in heart-rate. Appropriate changes of the order of 5 per cent were reported. Next DiCara and Miller (1968a), using the same basic design as the previous study, reported improved separation between the increase and decrease groups of about 20 per cent. The improvement was said to be due to the use of a shaping technique that allowed the animals to make gradual approximations to the final response. The studies were further extended to see if the same result could be produced by the use of negative reinforcement involving electric shocks applied to the tails of the rats. Attempts were also made to see if 'autonomic learning' displayed characteristics found in 'central nervous system learning' such as discrimination and generalization of the learned task to non-curarized states. Learning similarities were found and it was claimed that learning persisted over a period of three months.

These findings constituted powerful evidence that heart-rate could be conditioned in a curarized animal but Miller extended his studies to other autonomic systems (Miller and Banuazizi, 1968) and showed that learned responses were specific to the system being conditioned. Their experimental technique used heart-rate or intestinal contraction change as the operant or dependent variable. Other studies showed that rate of urine collection in the kidneys and flow of blood in the wall of the stomach (Carmona *et al.*, 1974) could be altered by use of operant conditioning techniques. DiCara and Miller (1968b) demonstrated that differential conditioning could be obtained between the two ear pinnae of rats. These findings, if confirmed, would serve as additional evidence to modify Cannon's statement that the sympathetic nervous system acted in an 'all or nothing way'. Unfortunately when this particular study was made, the authors restricted their recordings to the ear pinnae and did not also use their earlier technique of looking at flow rates in the vascular system of intestinal walls. This additional check would have provided evidence of what was happening to blood flow in a distant part of the vascular system.

PROBLEMS ASSOCIATED WITH ANIMAL EXPERIMENTS SHOWING AUTONOMIC CONDITIONING

The series of experiments carried out by Miller and his associates appeared to lend considerable support to the idea that it is possible to condition the autonomic nervous system using operant techniques. However, there are a number of problems associated with this research programme. For example, no detailed discussion appears to have been made by Miller and his colleagues about the action of the drug curare. This is important, for if the action of this drug is to interfere with the transmitter substance acetylcholine (Bovet *et al.*, 1959), then consideration of Figure 2.2, will show that curare should have some effect at the pre- to postsympathetic ganglionic junction. The effects of curare are not confined to the somatic system but also affect the autonomic nervous system. DiCara (1970) claimed that curare has no blocking effect on autonomic action but provided no evidence for a view that is contrary to most pharmacological evidence. Black (1967), in a very careful study using curarized dogs, reported that he could only obtain conditioning of heart-rate when the drug level used permitted small (detectable only with the use of a sensitive electromyogram) levels of muscular activity. It is now conceded that autonomic conditioning does not take place in the complete absence of muscle activity but further elaboration of this point is too detailed for consideration in this book. For additional information concerning the pharmacological properties of ganglia, interested readers should consult Eccles and Libet (1961). For an overview of the considerable problems associated with administering curare and the various techniques required to enable the paralysed animals to breathe artificially, see Hahn (1970) and Thornton (1971). Both of these areas can induce artefacts. Katkin and Murray (1968) have suggested that the types of reinforcement used in these studies (ICSS and electrical shock) are special classes of highly effective reinforcers that may bias results. In addition, Adolph (1967) has pointed out that in terms of autonomic function, the rat is a sympathetically dominant animal; this could tend to reduce its value as an effective animal for this type of research.

Finally, and more seriously, Miller (1973) has reported that it was proving difficult to replicate earlier studies where heart-rate separation between increase and decrease groups of 20 per cent were reported. Miller suggested, among other things, that the process of artificial ventilation can critically affect the results. For a review of animal research in this area, see Harris and Brady (1974). One moral to be drawn for future studies in animal research seems to be that too often in the past elegant experimental designs have taken precedent over having a comprehensive understanding of the physiological systems producing the behaviours.

HUMAN EXPERIMENTS INVOLVING ATTEMPTS TO CONDITION AUTONOMIC FUNCTIONS

As with the animal literature, earlier statements about the basic dichotomy of the learning process were accepted and helped to prevent the establishment of extensive and systematic research programmes. Mowrer (1938) attempted to use the galvanic skin response as an operant to avoid an electric shock, and Skinner and Delabarre (reported in Skinner, 1938) made a similar attempt using vasoconstriction. In both of these studies the findings were inconclusive but this did not prevent them being used as evidence against the possibilities for conditioning the autonomic nervous system. Kimmel (1967) reported that a search of the literature between the period 1920 and 1940 failed to reveal a single systematic study that could be used to support Skinner's view that it was impossible to condition the autonomic nervous system using operant procedures. However, by the late 1950s studies specifically designed to test Skinner's view, were beginning to appear in the literature. The major problem soon became very clear; it was, when you are using human subjects, can you be sure that they are not aware of the true nature of the learning task they have been asked to perform?

Clearly, in humans it is not possible to attempt to eliminate muscular movements by the use of paralysing drugs such as curare, although isolated studies have been reported in which human curare paralysis was induced in an attempt to examine this problem (Smith *et al.*, 1974; Birk *et al.*, 1966). In the Birk study the curarized subject showed effects of operant conditioning on his GSR despite being partially paralysed. Smith and his colleagues administered curare to a human subject over a period of 56 minutes and carried out a series of observations over a total period of four hours. After the drug was administered the subject progressively lost control of his muscles until he was only able to signal to his colleagues with his left eyebrow, administration of additional amounts of curare eliminated even this small muscle movement. The investigators concluded that at no time was the subject unaware of what was going on around him. Unfortunately, as pointed out by Malmo (1975), the study lacked the necessary controls needed to confirm this conclusion. No recordings were taken from any muscle endplates of the paralysed subject that might have revealed weak action potentials indicating information reaching the muscles from the brain. You will recall from the section dealing with the use of curare in animals, that it is now accepted that if the animal is paralysed to the extent that no action potentials can be detected then no learning takes place. In addition, the investigators failed to take objective measures of cognitive processes during the period the subject was paralysed. To merely ask the subject after the event allows him to reconstruct memories of what happened while he was paralysed. Presentation of unfamiliar stimuli to the paralysed subject for recall after the event would have helped to clarify the extent to which the subject was able to process information while his muscles were paralysed with curare.

Ideally, investigators concerned with cognitive processing during biofeedback studies should find the yokel featured in a *Punch* cartoon who is reported as saying (Spalding, 1965), 'Sometimes I sits and thinks and sometimes I just sits'.

Slowly the experimental problems and difficulties were revealed and solved, by 1967 Kimmel felt sufficiently confident to claim that tentative support had been found to show that it was possible to condition autonomic responses using operant techniques. However, as he pointed out, there were still a number of methodological points that prevented a firm answer to the question, can the autonomic nervous system be conditioned? A major problem concerned the use of the yoked design as an adequate control in experiments. When using a yoked control design, the experimenter places two subjects in identical situations but only one of them has control over the experimental contingencies. The yoked subject lacks control over the relationship between the stimulus and response.

For example, if the experimental subject is making a response to avoid an electric shock, his actions, or lack of them, are experienced by the yoked subject who is in an identical situation.

On the face of it, this control procedure appears to be completely satisfactory and adequate. But as Church (1964) has pointed out, even if two subjects are initially matched in terms of response-rates, the basic design contains a bias that results in the experimenter tending to get increased behavioural separation between his two subjects. The induced bias favours the conclusion that the subject controlling the contingency between the stimulus and response has learned the task while the yoked subject has not been able to learn the task.

The major theoretical and methodological problems have been reviewed by Katkin and Murray (1968) who made the critical distinction between conditioning and controlling autonomic functions. Conditioning experiments relating to the theoretical underpinning require a positive demonstration that autonomic conditioning has taken place in the absence of cognitive factors. However, control of autonomic functions in a clinical setting does not require elimination of cognitive factors. Indeed, it may be desirable that cognitive factors are present to facilitate the controlling processes.

Many of the earlier experiments in the area of biofeedback research were naive in that they appear to assume humans can be passive receivers of the experimenter's stimuli and quite unaware of the experimenter's contingencies. It is most doubtful if the idea of a passive experimental subject is valid when using lower animals, let alone man with his complex brain. In recent times, experiments have been made to explicitly examine cognitive variables used by the subjects. A good example of this type of investigation can be found in Stern (1967) in which a questionnaire was used in an attempt to identify the cognitive and somatic factors that had been used by subjects.

Recently, the *Journal of Psychophysiology* (Symposium, 1973) has reported a symposium concerned with cognitive factors involved in classical conditioning experiments using human subjects. The main conclusion of this symposium was that if attempts are being made to condition or alter autonomic activity in humans, then the more aware the subject is of the exact nature of the changes required, the better are the chances of success. Learning studies using masking tasks designed to conceal the true nature of changes required, generally lead to inferior performances. This finding was supported by the fact that detailed questionnaires, given for the purpose of probing for cognitive factors after a learning task, show that subjects who have the highest 'awareness' levels of the nature of the task also show the best performances. From these studies we conclude that in the clinical setting, attempts should not be made to conceal the nature of the task from the patient. This position is also favoured by recent learning theory (Rescorla, 1967) attempts to reintroduce cognitive factors into the Pavlovian learning situation. The theoretical position taken by Rescorla holds that the organism acts as a contingency analyser rather than a passive producer of responses. Furedy (1971, 1973) has discussed this theoretical aspect in some detail and produced some experimental evidence against it. In a later paper Furedy presents evidence which, he argues, supports the view that it would be rash to assume that autonomic conditioning cannot take place in the absence of cognitive factors.

It is possible to obtain changes in autonomic activity without awareness? In a recent publication Dawson and Furedy (1976) state that it is not possible. These authors suggest that the autonomic nervous system cannot be conditioned in the absence of awareness; although, by itself, the presence of awareness in no way guarantees satisfactory conditioning. This view holds that awareness is related to conditioning in an all-or-none fashion. However, there are some earlier reports which run counter to the above argument. For example, Hilgard and Humphreys (1938a), 1938b) carried out an experiment in which they conditioned subjects who were later tested for retention of the task and awareness of the cognitive factors they had used. It was found that although the subjects were unaware or had forgotten the cognitive factors they used in the task, they retained the learned performance. This is an important experiment that requires to be repeated because it may point to an essential separation between the autonomic activity, which is being used as the behavioural performance or dependent variable (located in areas of the hypothalamus), and the cognitive components (located in the cortical and frontal lobes).

This hypothalamic interaction with the frontal lobes revolves around the fact that the main physiological mechanisms required for activity of the autonomic nervous system appear to be finally mediated via the hypothalamus. Certain reflex behaviours may be triggered without too great an involvement of the higher cortical centres. However, it seems likely that autonomic learning will always be facilitated by active involvement of higher cognitive centres. Any factors that serve to facilitate links between the 'policy making' areas of the frontal lobes and the 'executive' areas of hypothalamus will aid autonomic learning. This view inclines away from Dawson and Furedy's all-or-none view and supports the idea that awareness is related to conditioning in a continuously functional relationship. Detailed consideration of this point will be left until emotion has been discussed but basically it will be argued that the brain has awareness of autonomic function. Usually awareness of autonomic function is low because often information processing areas take precedence over the more routine aspects of internal functioning.

One area that has received little attention is autonomic conditioning in early childhood. To take an example, a child may fall and hurt himself sufficiently badly to produce the response of crying, this is largely a reflex response. The important point for the cognitive position is that a child may construe that the sympathy and attention received relates to the level of crying after the accident. The crying response is then generalized and subsequently in every minor accident, tears are used to elicit sympathy from adults. Later the child, especially if he is a boy, may need to learn that he must suppress his tears. In the older child it is the proverbial 'stiff upper lip' which gains admiration and sympathy. This general area of autonomic conditioning in young children appears to have been neglected but obviously it could be of crucial important. In relation to later function or malfunction, the early years of life may be essential for setting the operating levels for later somatic/autonomic interactions. Lipton *et al.* (1965) have reported some Russian work in which it has been claimed that immature babies show greater dissociation between the autonomic and somatic components of their response to a conditioned response.

To large degrees intergration betwen the somatic and autonomic systems must be crucially dependent upon individual early learning and conditioning. By the time an adult becomes a subject in a psychophysiological experiment, the physiological as well as the psychological idiosyncrasies are formed and set. Moreover, the problem for the experimenter is further confused by the fact that certain subjects will have a predominantly visual memory while others may rely on verbal mechanisms. Little wonder then that it is difficult to find psycho-physiological and biofeedback experiments showing clear cut results.

The emphasis in research into biofeedback topics has now switched from concern with behavioural models and methods to concern about the basic underlying physiological mechanisms (Shapiro, 1977), although there is still a considerable way to go before we have sufficient knowledge of the fundamental details to warrant some of the previous claims. Obrist (1976) has pointed out that a far greater sophistication is required concerning the biological significance of the psychophysiological systems being investigated. The 'oversell' of biofeedback in relation to its potential in the clinical fields has led to a sense of disillusionment that is apparent in recent publications and conferences (Beatty and Legewie, 1977; Schwartz and Beatty, 1977). The disillusionment appears to relate in most cases to lack of fundamental knowledge. The cognitive mediation hypothesis has not been unequivocally demonstrated either way and few of the positive results claimed for humans appear to generalize outside the confines of the research laboratory. The lack of fundamental knowledge means that in most cases even if an autonomic system is successfully conditioned, it is not possible to satisfy scientific criteria by saying how and why the technique was successful. Garcia and Ervin (1968) have suggested that there are quantitative as well as qualitative differences between the somatic and the autonomic nervous systems that require to be carefully examined and analysed. In their study Garcia and Ervin used an avoidance conditioning technique that involved subjecting rats to unavoidable X-rays at the same time as they were eating, and found that aversive conditioning could take place over very long time delays. In normal conditioning experiments using positive reinforcement, it is found that very short delays must occur between the stimulus and the response if learning is to take place. Garcia and Ervin demonstrated that the irradiation sickness although occurring many hours after the ingestion of food was still associated with the food. Their finding led them to suggest that visceral receptors are different from the teleoreceptors of the brain in that very long delays may be interpolated between the stimulus and the response. Garcia and Rusiniak (1977) have questioned the use of traditional learning techniques to condition discriminations of the autonomic system. Garcia and Rusiniak point out that the continued use of short delays between

the stimulus and the response might be counterproductive, also the use of exteroceptive stimuli to signal cognitive information to the subject about changes in
his autonomic system might be counterproductive tending to interrupt induced
states of relaxation. Garcia suggests that it might be more appropriate to use
hedonic or pleasurable kinaesthetic changes as the reinforcement arguing that
these appear to be more conducive to autonomic processes. It would certainly be
of value to use a variety of methods using both teleoceptive and visceral
receptors to compare the relative efficiency of conditioning obtained by
different types of receptors. Garcia's position may not be too popular at present
but it is true to say that learning theorists' techniques, worked out in the laboratory to examine learning processes of animals, appear to have limited validity in
the biofeedback laboratory and it would be of value to widen our approach. A
wider approach has also been suggested by Green and Green (1977) who
advocate the increased use of relaxation techniques.

Lack of fundamental knowledge about the action of the autonomic nervous
system is revealed in a comment made by Katkin and Murray (1968) in an other-
wise sound review when they say that the autonomic nervous system does not
interact with the outside world. This statement is incorrect as it stands and
requires qualification because the autonomic nervous system does interact with
the external world. Undoubtedly the nature of the interaction between the
autonomic system and the external world is very different from the interaction
of the exteroceptive senses of vision and hearing but, as illustrated in Chapter 3,
changes in the ambient temperature cause autonomic changes. The digestion of
food inside the intestine is an interaction between the body and the outside
world because the digestive tract is external to the body; digestive processes only
become internal once the nutrients have passed through the walls of the
intestine. This may be a small example but it does highlight the lack of
knowledge and serves to perpetuate the myth that there are two separate and
quite distinct nervous systems within the body. There are differences but then so
there are between the senses of hearing and vision. What we need to know is how
and why they are different. Johnston (1977), in the area of cardiovascular
feedback, has failed to find more than limited benefits using a motor-skills
analogy but clearly the motor-skills analogy, and its possible relationship to
autonomic conditioning, requires anlaysis in greater depth and for a wide range
of autonomic functions. A good example of the need for this detailed knowledge
can be found in a study by Little and Zahn (1974) who examined changes in
mood and autonomic function during the menstrual cycle of a group of females.
They found that although breathing and heart-rate increased during the luteal
phase of the menstrual cycle skin conductance fell. At the present time there is
no explanation for their paradoxical finding which, incidently, also runs
counter to the concept of general arousal.

One important question that still remains to be discussed in detail is, to what
degree is autonomic perception possible? Indeed, is autonomic perception
possible at all? Detailed information in the literature is rare and views range
from 'not possible' to 'certainly possible'. The historical details of this problem

have been considered at a number of points in the text but for an interesting and more complete overview the reader is referred to a review by Kimmel (1974) which, with apologies to Kimmel (1974) and Razran (1972), can be colourfully subtitled; how psychology lost and rediscovered its 'guts'.

Mandler (1975) takes the more commonly held view about visceral perception in maintaining that visceral perception is very low. He points out that the paucity of internal receptors that can pick up changes in autonomic function indicates that the vital role of visceral perception is likely to be arousal and that fine-grain differences in autonomic awareness are unlikely to be sensed by humans. Mandler, basing his case upon results he obtained from his autonomic perception questionnaire (APQ), argues that the relationship between autonomic perception and visceral discharge while positive is relatively weak (Mandler *et al.*, 1958). However, it has been pointed out that a number of studies have shown that the APQ tends to measure the vocabulary of the words that subjects use to describe their autonomic function rather than a true measure of their ability to detect internal visceral changes. In fact, later investigators have shown that subjects who score high on the APQ tend to overestimate their autonomic activity while subjects who score low tend to underestimate their autonomic activity (Brener, 1977).

Brener (1977), in a timely and pertinent review on the problem of visceral perception, raises a number of interesting questions and points out that attention is usually directed towards the goal of our actions rather than reflecting upon the pattern of the interacting processes that are producing our actions. Indeed, we must not forget that our need to communicate with others has resulted in emphasis upon cognitive factors and this fact is reflected in the education we give children. There have been long and learned philosophical debates about the nature of external perceptions and humans have come to learn how to analyse and explain them to others. MacLean (1949) points out that in children and preliterate peoples there is often a failure to discriminate between internal and external perceptions; he gives as an example a child looking at a leaf and saying, 'it tastes green'. Visceral feelings are blended and fused with what an individual sees or hears and this is done in such a way that the outside world is incorporated with the internal world.

At least on the face of it, there are also some good examples of visceral perception. Stunkard and Koch (1964) in a gastric perception experiment showed that people of normal weight tended to report hunger sensations that were correlated with gastric motility whereas obese subjects tended to report hunger sensations that were independent of their gastric motility. Schwartz (1975) has reported that subjects can learn to raise their blood pressure while lowering their heart-rate; in addition, they can, to a lesser degree, learn the converse task of decreasing blood pressure and increasing heart-rate. It should be emphasized that these two operations are carried out at the same time. The secret of the method according to Schwartz lies in teaching subjects the separate elements of the overall patterning that is required before asking them to combine the elements into a single response. In this respect Schwartz is more favourably inclined towards the sensorimotor skills analogy than Johnston (1977). Brener (1977) in conclusion states that the idea of the visceral afferent system being substantially different from the exteroceptive or proprioceptive systems is severely compromised by an overwhelming body of evidence.

The finding of cognitive factors involved in autonomic activity should not be surprising. Once the evolving brain had progressed beyond the level of the spinal cord and lower brain stem, it did not leave control of lower reflex behaviour behind, but at each stage carried forward control and integrated components into the new brain structures and their functions. Normally we do not require extensive control over autonomic activities for *Homo sapiens* is an information processor of the external world and refinement of his cognitive functions has resulted in a relative decline of control over internal activities. For humans it is unadaptive to spend considerable amounts of their time consciously manipulating their internal environment. At this point, you may recall the desert lizard who needs to spend so much of his life behaviourally manipulating his body temperature. However, internal activities of humans are not carried on in the total absence of awareness by the higher centres, the point is that the level of 'awareness' or monitoring is only required to be routinely low. At times of malfunction this level of awareness may increase markedly. In fact, as many have discovered during a 'hangover', the autonomic nervous system can come to dominate the cognitive functions almost completely. Premenstrual or menstrual tensions seem another good example of where internal visceral events colour or dominate cognition.

We can illustrate this problem of 'awareness' by the brain by pointing to the action of walking which is often assumed to be automatic. If we carry out a demonstration to see if this is the case by asking a person who is walking along a flat and wide corridor free from obstructions, to perform a reasonably simple multiplication sum in his head, invariably the person will stop walking. This illustrates the fact that when the attention of the brain is directed towards a cognitive task it may suspend routine activities. In the situation just described, we show that walking is an overlearned task, requiring a low level of 'awareness' or monitoring by the brain. In addition if we observe the acquisition of the walking skill in young children, we find that it is learned slowly and sometimes painfully over a period of time that can be protracted. Consider then the subtle level of control required in most autonomic functions that are not learned consciously because they are largely reflex responses. Normally we do not require or need extensive control over autonomic functions but the brain has provisions for such a possibility should it become necessary.

In terms of evolution, brain stereotypy has usually meant extinction. The giant reptiles who occupied this planet for long periods had minute brains that were more than adequate for controlling their huge bodies, until their environment began to change and animals with more flexible cognitive brains evolved. From that moment the giant reptiles were doomed to extinction. They had evolved for a narrow ecological band and perhaps one of their problems was the lack of modifiable internal processes to enable them to cope with more than moderate changes in their environment.

Despite these facts textbooks and monographs are still being written about the psychophysiology of the frontal lobes that contain no discussion or reference to autonomic function (Pribram and Luria, 1973). Can we argue that there are internal physiological processes that do not require cognitive processing? The position favoured in this book is to see the problem of cognitive involvement in physiobiological events as ranging along a continuum of brain awareness from maximum to minimum with certain reflexes (e.g. the clasp-knife reflex of the tendons that serves to prevent excessive loading on the bones by muscles), where the level of involvement by the brain is very small if not negligible.

Up to this point in the chapter we have refrained from introducing the concept of emotion believing that it was better to keep the concepts of biofeedback and emotion separate. However, at this point we must turn to discuss the relationship between the autonomic nervous system and emotion. As Brener (1977) has pointed out, metaphors concerning the autonomic nervous system have long been a favourite way of communicating ideas about emotion and Watson (1928), the archbehaviourist himself, wrote:

'We may earn our bread with the striped muscles but we gain our happiness or lose it with the kind of behaviour our unstriped muscles or guts lead us into.'

THE AUTONOMIC NERVOUS SYSTEM AND EMOTION

In recent years there has been a resurgence of interest in peripheral theories of emotion (Goldstein, 1968; Fehr and Stern, 1970) that appears, in part, to have arisen from the failure to find adequate central nervous system theories of emotion. The literature on the topic of emotion is voluminous and definitions abound for all tastes. It is beyond the scope of this book to attempt a comprehensive discussion of emotion. Overviews of the four major research symposia of emotion can be found in Murchison (1928), Reymert (1967), Arnold (1970), and Levi (1975). Good introductory texts have been written by Strongman (1973) and Candland et al. (1977). Comprehensive overviews of the problems confronting a psychological theory of emotion will be obtained from Plutchik (1962) and Candland (1977).

Before Darwin (1872) it was assumed that emotion was an element of consciousness but he proposed the novel idea that it was possible to consider emotion as an organic process having adaptive value for survival. Darwin opened the way for later studies which attempted to find the physiological structures responsible for emotion and also pointed to the important distinction between emotional *feeling* and emotional *behaviour* but it was a considerable time before this distinction was fully understood. Emotional feeling or experience may be purely subjective and not be obvious to the observer. Emotional behaviour, by definition, is always obvious to the observer. Of course, you might ask: 'Is it not possible that the observed "emotional behaviour" is merely a performance (a conditioned response) rather than the reflection of a genuine emotion?' The answer is, 'Yes it is possible,' and therein lies the crux of the difficulty for the experimental psychologist studying emotion. Undue emphasis upon observable emotional behaviour has resulted in counts of rodent droppings (an undeniable quantitative measure) being used as an index of emotion on one hand (Tobach, 1969) and virtual silence about human emotion and its role in cognitive functions. As we shall see, the confusion between the emotional experience / behaviour categories has led to a great deal of muddled thinking. But this is taking us too far ahead, so let us retrace our steps back to the last century and return to develop this theme later.

THE JAMES-LANGE THEORY OF EMOTION

Darwin placed emotion firmly into a physiological mould but it was not until the end of the century that three workers (James, 1884; Lange, 1887; Sutherland, 1898) independently proposed a theory that could be experimentally tested. Of the three, James clung most firmly to the former introspectionist position concerning emotion and as a consequence, his particular theory is difficult if not impossible to test. Original accounts of the theories of both James and Lange have been reproduced in a monograph (Dunlap, 1967). Sutherland was an Australian who formulated his version of an organic theory of emotion before he was aware of the James-Lange theory.

Lange, although assumed by most to be a pale imitation of James, was very concerned to closely relate emotion to a bodily function and from his experience as a medical practitioner, he chose the sympathetic nervous system to control emotion. In a passage that clearly states his position and, incidentally, shows that he was well able to turn the hyperbole with James, he wrote:

'We owe all the emotional side of our mental life; our joys and sorrows, our happy and unhappy hours, to our vasomotor system. If the impressions which fall upon our senses did not possess the power of stimulating it, we would wander through life unsympathetic and passionless, all impressions of the outer world would only enrich our experience, increase our knowledge, but would give us neither care nor fear.'

Few research workers would care to support such an extreme position today but we can recognize that this passage would provide a feasible research programme. Dunlap (1967) has suggested that psychologists would profit more from following Lange rather than James and, in fact, Wenger (1950) used Lange's theory to suggest how one might begin to programme the dimension of emotion into a robot. James's most celebrated analogy concerning this theory, is of a man suddenly sighting a bear and turning to flee before realizing that he was afraid. James argued that the action of escape came before the emotion of fear was experienced. Unfortunately, he never dealt with the situation where the person 'froze with fear'.

The James-Lange theory is the most famous of the peripheral theories of emotion and it is reminiscent of the ancient view that sensation arose from the backwash of animal spirits from the peripheral parts of the body. In no small part, due to the fact that the theory was not an introspectionist's account it came under immediate attack from all sides but we shall confine our attention to criticisms based upon experimental work.

ANIMAL STUDIES RELATING TO THE JAMES-LANGE THEORY

Functional studies of the role of the autonomic nervous system and emotion have mainly been confined to analysing the role of the sympathetic nervous system which, as Cannon showed, can successfully be removed in an animal confined to the laboratory. The parasympathetic nervous system is essential to life and it cannot be removed or blocked for extended periods. Interference of the sympathetic nervous system has been by three main methods: (1) Surgical means, these involve three techniques: (a) transection of the spinal cord; (b) removal of the chains of paravertebral ganglia alongside the vertebral column (sympathectomy); (c) denervation or removal of the adrenal medulla. (2) Pharmacological means, using active substances that have a general or a specific effect on the biochemical processes of the autonomic nervous system. (3) More recently it has become possible to induce hypotrophy in developing sympathetic ganglia by injecting a specific antiserum into neonatal animals (immuno-sympathectomy).

Historically, the earliest attempt to test the theory experimentally was made by Sherrington (1900) who transected the spines of a limited number of dogs and tested them in naturalistic settings after they had recovered from the surgery. His tests were ingenious and they included observing reactions of the dogs to normal handlers and strangers. Sherrington presented the dogs with cooked dogflesh and observed that they showed repugnance which he interpreted as emotional reaction against eating flesh of their own kind. However, he failed to carry out a control test which would have been to present his dogs with other unusual and strange meats. For example, would his dogs have shown a similar repugnance to goat's meat? Sherrington also failed to have neutral observers rating his dogs in order to eliminate experimenter bias. Had he carried out these appropriate controls, he might not have concluded that his dogs showed no marked deficits in their emotion. Missing the finer nuances of the James-Lange theory, Sherrington went beyond observations of the dogs' emotional behaviour and drew conclusions about their emotional feelings that were unwarranted.

Cannon (1929, 1931) also attempted to experimentally disprove the James-Lange theory by using sympathectomized cats from which he had removed the paravertebral ganglia (Cannon et al., 1927). Like Sherrington before him, Cannon used seminaturalistic tests and exposed his cats to strange and fierce dogs. As with the Sherrington study, Cannon concluded that, subjectively, he was unable to detect any emotional deficiencies in his experimental animals (Cannon et al., 1929). But in this study also, there is a crucial control condition missing for it is not possible to decide if Cannon's sympathectomized cats were producing a conditioned response to the sight of a fierce dog. What Cannon's cats may have been producing is appropriate emotional behaviour. Had it been possible for these animals to report on their introspections to the sight of a strange dog, they might have indicated that they did feel less fear than formerly would have been the case. Both of these early studies made the error of making assumptions about emotional experience from subjective assessment of emotional behaviour. In addition, Cannon's sympathectomy technique involves a number of problems concerning the removal of the sympathetic nervous system. As pointed out in Chapter 2 one of the main problems is super-sensitivity, and careful checks would be required before any final statements could be made.

The conclusion from these studies was that the James-Lange theory was well and truly defunct but even as the last rites were being held (Cannon, 1929, 1931), the theory was coming back in a modified form in the area of learning theory as the two-factor theory. The two-factor theory of learning was dealt with at length earlier in this chapter and you may recall that Mowrer proposed the reason why animals in an avoidance situation did not extinguish their responses was that the warning signal (CS) elicited autonomic responses that had formerly been elicited by the shock (UnCS). Mowrer and Keehn (1958) argued that intertrial-crossings made by animals in a shuttle-box test of escape and avoidance conditioning came about because environmental stimuli triggered off autonomic responses. By 1950 animal learning theorists were fully aware of the distinction between emotional feeling and emotional behaviour and studies began to concentrate on the latter.

Mowrer's two-factor hypothesis was the basis for the most complete study of peripheral factors in emotional behaviour. (Solomon and Wynne, 1950, 1953). Wynne and Solomon (1955) used a sympathectomy procedure that was very complete and consisted of removal of both sets of the paravertebral ganglia and also the greater splanchnic nerves to the adrenal glands, see Figures 2.1 and 2.6. In a number of animals they also blocked the parasympathetic vagus nerve by resecting it on one side of the neck and pharamacologically blocking it for short periods. By this almost complete treatment, they were able to assess the relative contribution of the autonomic nervous system in certain tests. Solomon and Wynne's test of emotional behaviour consisted of a shuttle-box where the dogs were required to leap over a central barrier in order to escape or avoid a signalled 12 mA electric shock. Their analysis of the results consisted of categorizing various types of behaviour seen. In general, sympathectomized dogs were said to be quieter than normals except when in the test condition where they tended to display greater variability. However, no overall uniformity was found in the types of behaviour seen. One of the ways a less emotional animal might show itself in terms of behaviour is by showing rapid extinction of a learned avoidance response and the authors reported that several of the sympathectomized dogs showed spontaneous extinction. Wynne and Solomon also emphasized the fact that the experimental animals were slower to make the first avoidance response but eventually they all learned to avoid successfully. The final conclusion from this study was that under the conditions of their experiment, the autonomic nervous system played a significant but not essential role. One complicating factor in the three studies reported above is that the use of adult animals might have resulted in genuine differences between the experimental and control animals being masked by previously conditioned responses. An adult animal would have extensive experience of the function of the autonomic nervous system during his life-span so that, in the test situation after surgery, he may have produced 'emotional behaviour' to environmental and situation cues. In fact, Solomon and Wynne reported that their dogs were taken from the local stray dog compound. One might expect that these animals were familiar with aversive situations even though they are unlikely to have had previous experience of electric shock. The point being made is that aversive situations, both in real life and the experimental psychologist's laboratory, usually call for activity to successfully overcome them. One way to examine such a possible conditioned response interaction would be to remove the sympathetic nervous system before the animal had any extensive experience of its action. Levi-Montalcini and Angeletti (1966) have summarized a series of experiments that relate to an immunological technique for removing a major part of the sympathetic nervous system within a short period of birth. The technique, called immunosympathectomy, uses an antiserum prepared from the Nerve Growth Factor isolated from the salivary glands of male mice. Hypotrophy of sympathetic ganglia is induced in neonatal animals by daily injecting Nerve Growth Factor antiserum for the first few days after birth. Summaries of research studies are reported in two publications (Steiner and Schonbaum, 1972; Zaimis, 1972).

Two main series of behavioural experiments have been reported using immunosympathectomized mice and interestingly they come to different conclusions. Wenzel (1972) summarising a large number of experiments using these mice has concluded that immunosympathectomized mice are defective in their ability to mediate avoidance situations. This conclusion supports the two-factor theory of learning and is broadly congruent with the earlier findings of Solomon and Wynne. However, Van Toller and Tarpy (1974) have disagreed with this conclusion and have argued that unequivocal behavioural separation between immunosympathectomized and control litter-mate animals has not been shown in an avoidance situation. Further, they argue that it is more parsimonious to look at specific roles of the sympathetic nervous system rather than invoke a general behavioural paradigm involving mediation of aversive situations by the sympathetic nervous system. In support of their arguments, they subjected immunosympathectomized mice and control litter-mates to cold stress for 2 hours prior to running them in an escape avoidance task involving an alleyway (Van Toller and Tarpy, 1972). The results of this experiment showed that the immunosympathectomized mice were significantly and progressively inferior over days. It is not essential to detail all the points of the argument in this chapter and interested readers are referred to the references given.

Of the original techniques quoted at the beginning of this section, we have still to consider (c) removal of blocking of the adrenal gland and the use of (2) pharmacologically active substances. In both of these areas there are numerous studies in which the distinction between emotional behaviour and emotional experience has not been made clear; however, both types of studies can be summarized fairly succinctly.

Caldwell (1962) made an attempt to relate directly the function of the adrenal medullary cells to emotional states. He found that demedullated mice froze, squealed, and groomed less often than normals, while being trained in a conditioned avoidance technique. These findings, together with the fact of longer latencies and increased exploration, were taken as evidence that the experimental mice were less emotional than the control animals. Levine and Soliday (1962) using rats, reported a small effect on a conditioned avoidance task following adrenal medullation. Conversely, Moyer and Bunnell (1959, 1960) were unable to find any significant effects between demedullated and control animals. Pare (1969) obtained negative results in rats and suggested that this particular species is not 'adrenal dependent'. Taken overall, the results are negative and, as pointed out in Chapter 2, there are considerable problems relating to adrenal demedullation. For example, the role in demedullated animals of extrachromaffin cells in the abdominal region producing catecholamines.

The use of pharmacological agents to block autonomic nervous system function is complicated by numerous factors, ranging from uneven effectiveness of the agent over time to side effects that may interfere with the particular systems being analysed. As Weiskrantz (1968) has pointed out, if specific effects are sought, then parametric studies are crucial if valid conclusions are to be made but all too often these are not made.

Auld (1951) reported a study in which he used the drug tetra-ethylammonium chloride (TEA) to block transmission of efferent impulses through autonomic ganglia. Using a straight alleyway task he analysed avoidance responses of normal and drugged rats. The drugged group had slower running times suggesting that the underlying fear response was at least partially reduced. Auld's conclusions were questioned by Brady (1953), who studied the effects of TEA on a number of tasks which included wheel running, lever pressing, and running in an alleyway. Brady's study showed that small injections of TEA induced muscular asthenia that resulted in a considerable interference of the motor performance in his rats. Davitz (1953) also found that rats in which a freezing response had been induced, displayed less activity on alternate extinction trials using TEA. These findings are supported by an earlier observation (Gruhzit *et al.*, 1948) which studied the effects of acute and chronic administration of TEA in rodents. Finally, Arbit (1957, 1958) reported that the use of a second blocking compound, hexamethonium, which has less effect on skeletal muscles, failed to produce a significant difference between the drug and control groups which he found present when he tested his animals with TEA. Thus, it would appear doubtful if drug studies up to the present time have been sufficiently free of side effects to make a meaningful contribution to our understanding of the role of the autonomic nervous system in mediating emotional states.

This has been rather a long subsection of the book but it should be realized that the main experimental evidence for and against what we might call the physiological structures involved in the James-Lange theory of emotion has come from animal studies. All too often these studies have been confused and have not tested what they set out to test. This arises because workers often do not seem to understand the autonomic nervous system and its many roles. Added to this are the complications that arise from the experimental procedures themsleves. Experimenters should begin to break away from the now-traditional methods of testing emotional behaviour and emotional experience. Traditional methods appear to be more related to tests of stress rather than testing the natural emotional behaviours of the species being tested.

Psychology like the other sciences operates best in situations where it can define its terms and circumscribe its area of interest and the main paradigm used in learning theory has been the stimulus response or S-R bond. However, emotion has not proved to be amenable to this type of treatment, it will not break down into discrete pieces that can be meaningfully handled in the laboratory. Mandler (1975) feels that the two-factor theory application is limited to young children and animals; interestingly, in both cases we find a lack of cognitive ability to make complex analyses. Mandler holds that the two-factor theory is not a general theory of avoidance behaviour but a limited example where the reaction to the stimulus is limited and stereotyped. By moving away from consideration of emotional experience and concentrating upon emotional behaviour psychologists have limited themselves unduly. A moment's consideration reveals that the transient emotions used in most laboratory studies have little in common with the chronic emotions that the clinical psychologist confronts daily.

HUMAN STUDIES RELATING TO THE JAMES-LANGE THEORY

The problem in this part reverses and becomes not one of what to include as representative of the various limits of research but of searching for the limited number of studies that have been carried out. However, there are a few that directly relate to the James-Lange theory of emotion. The first clinical report was made by Dana (1921) who, arguing against the James-Lange theory, presented clinical evidence from a middle aged woman who transected her spine in a fall from a horse. Following her unfortunate accident, the woman lived for a year and Dana argued that during this period he was unable to detect any loss of emotion. One must presume that he was talking about emotional behaviour and saying that her behaviour was always appropriate for the situation, but he is never clear on this point in his short account. If Dana's statement is about the appropriateness of his patient's social responses then it raises the problem of conditioned emotional responses or perhaps conditioned social responses. We can surely presume that a middle-aged woman would have learned the full complement of emotional behaviour. If a friend brought her a gift or she came to hear of some sad news, she might well respond with the appropriate emotional behaviour, particularly if she was in a social situation with other people present. Unfortunately Dana's study appears to have been taken as conclusive evidence that the sympathetic nervous system did not play any extensive role in mediating emotion in humans as no further work was reported for several decades, until Hohmann (1966) reported a study made on paraplegic patients suffering from spinal transections at various levels. Hohmann claimed that the paraplegics, whose incapacity followed an accident, did experience a reduction in the intensity of their emotional feeling. Paradoxically, these same patients had been tested earlier by McKilligott (1959) who reported that they did not suffer from a diminution of their emotional feelings following their accidents. How did this reversal of the experimental finding come about? Hohmann argues that his finding of a reduction in emotional intensity is the correct one as he himself is a paraplegic and he was able to identify closely with the patients who were prepared to admit to a reduction of emotion to him. In the case of a normal experimenter (McKilligott), the patients were concerned to preserve a picture of what they considered to be a normal person's emotional behaviour and not reveal any emotional defects.

This finding seems to reopen the emotional behaviour / experience controversy in the clinical field. Whatever the final conclusion, it seems undoubtedly true that the patients produced the kind of information they 'knew', or thought they 'knew', the experimenter required. Most of the information in the two studies was collected by questionnaires and it is difficult to determine if sufficient attention was paid to the passage of time as a critical variable. What is a patient who suffered an accident 10 years before, answering when he replies 'yes' to the question: 'Do you experience less emotion now than you did before your accident?' We would all claim some diminution of our emotions over such a long period of time. It seems possible that the patients were answering in terms of a time relationship rather than one related to the fact that parts of their sympathetic nervous system had been separated from their central nervous system. In support of Hohmann's position concerning paraplegics, Hester (1971) has reported lower proprioceptive feedback in 20 patients with complete tran-sections of cervical and upper thoracic sections of their spinal cord. Fuhrer and Kilbey (1967) found lowered skin resistance responses in transected patients. Clearly, in this type of study there are a number of problems that require very careful experimental designs if firm conclusions are to be drawn. In the clinical field we again run into problems that have not been resolved and we find ourselves unable to reach any firm conclusion about the exact role of the autonomic nervous system in emotion.

Is it possible to draw any conclusions about the patterning of autonomic responses in different emotions? You will recall that Ax and Funkenstein had earlier claimed differences for fear and anger states. Sternbach (1962) measured six autonomic reactions in young children viewing Walt Disney's film *Bambi*. He found a lack of consistency in the autonomic response patterns displayed by the children and also in the direction when correlations were made with subjective judgements made by the children as to which scenes they thought were the 'happiest' or the 'saddest'. In conclusion, Sternbach was not able to do more than suggest that the technique of using films to study autonomic response patterning was potentially useful.

In an extensive study Averill (1969) also used the film-viewing technique as a method of studying autonomic response patterns. His subjects were divided into three groups and each group assigned to view a sad, funny, or neutral film. Subjects viewed the films in isolation and eight autonomically innervated responses were measured; these included finger and face skin temperature. A complete analysis of the data was made but the main analysis concentrated upon the 'sad' and 'funny' groups. Averill found that both these groups showed a tendency towards sympathetic activity, with both groups showing prominent electrodermal changes. The group viewing the sad film were characterized by cardiovascular changes and the group viewing the funny film were characterized by respiratory changes. Averill (1968) has argued that sadness and grief are catabolic states. (This may relate to the heightened isometric muscle tension shown in these states.)

Before we leave this part there is an interesting clinical complaint that involves an interaction between the sympathetic nervous system and emotion. In a doctoral thesis presented in 1862 Maurice Reynaud described the symptoms of a disease that was later to bear his name. Reynaud's disease is a complaint in which sufferers produce severe vasoconstriction of their peripheral blood vessels in response to cold. The disease is found mainly in females but it occurs in males whose occupations involve the prolonged use of vibrating power tools (Johnson and Spalding, 1974). The main interest for this book is that patients produce their vasoconstrictive overreaction to emotional stimuli. Mittelmann and Wolff (1939), in a paper entitled 'Affective states and skin temperature', provided evidence for the clinical view that emotional stimuli could produce the symptoms of this disease in sufferers. The series of studies carried out by Mittelmann and Wolff have been neatly summarized by Malmo (1975). Mittelmann and Wolff found that in an interview situation where emotional reactions were evoked, their patients showed severe levels of vasoconstriction that surpassed levels produced in response to cold stress. The type of emotional stimuli that produced the most severe response were those evoking a hostile or attack response. Patients' vasoconstrictive overreaction was increased if the ambient temperature was lower than normal.

In addition to the passage reproduced at the start of this chapter, Lange, with reference to vasoconstriction, quoted Shakespeare's 'Go prick thy face and over-red thy fear, thou lily-livered boy' (*Macbeth*, Act V, Scene 3). Unfortunately, there appears to be no record of Lange and Reynaud meeting or communicating. A collaboration between them might have proved to be very fruitful for the study of the interaction between emotion and the sympathetic nervous system. Stanley-Jones (1966) has also suggested that human emotion is rooted in the mammalian defence against hot and cold temperature changes.

CENTRAL NERVOUS SYSTEM THEORIES OF EMOTION

Before we consider modern theories of emotion that involve models of the autonomic nervous system we should summarize the central nervous system theories of emotion that arose largely because of Cannon's criticism of the James-Lange theory of emotion. Cannon argued that the structure of emotion would not be found in the autonomic nervous system due to five main reasons. The first concerned his studies of emotion using sympathectomized cats. He claimed to have found emotional states in his experimental animals lacking a sympathetic nervous system. As we have already mentioned, he could not be sure about this finding because he confounded emotional behaviour and emotional experience and it is possible that his adult cats were producing a conditioned emotional response to his test situation. Second, he held that it is difficult if not impossible to specify specific visceral changes that should exist in different emotions. As we shall see in a later section this specificity may not be a necessary requirement. Cannon's third point related to the study reported by Marañon (1924) who showed that injections of adrenaline given to subjects failed to produce emotion. Cannon assumed that adrenaline was the transmitter substance for the sympathetic nervous system and as we have already pointed out this was an incorrect assumption. In any case the Marañon's subjects did report emotion or feelings that he called the 'as if' emotion. We will consider this 'as if' emotion later. The fourth reason related to the relatively slow speed of nervous reaction found in the autonomic nervous system. Cannon's fifth and final point related to the diffuse nature of the autonomic nervous system which he felt seemed too unlikely a structure to mediate the fine differences found in emotion. James was too wily a philosopher to be disposed of so easily and Cannon, despite the intuitive appeal of a theory of emotion involving the brain, never really succeeded in eliminating the James-Lange theory of emotion which also contains a certain intuitive appeal.

The Cannon-Bard theory of emotion as it came to be called proposed a central theory that involved the thalamus. As originally postulated (Cannon, 1929) the relay of impulses from the thalamus to the cortex resulted in awareness of a stimulus but no emotional colouring. However, if the arrival of the impulses from the thalamus resulted in thalamic release from cortical inhibition then emotion was experienced. This was the 'emotional quale' that when added to a perception gave it the peculiar quality of emotion. Cannon stated that under certain circumstances thalamic processes of emotion could be directly aroused by afferent impulses from the receptors. It differed from the James-Lange theory that involved the thalamus. As originally postulated (Cannon, 1929) the arise as the result of thalamic emotional processes rather than being caused by the emotional feelings. Bard (1934) later added that the thalamic process by itself was inadequate and a signal to the cortex was required if the emotional quality was to be added to perception.

Bard and Mountcastle (1948) showed that if the brain of a cat was transected at the level of the thalamus then the animal was still able to display emotion. The emotion, or emotional behaviour as it should have been called, was peculiar and in recognition of this fact they called it 'sham' rage. The most marked characteristic of 'sham' rage was that it switched 'on' and 'off' with great rapidity. For example, if one of their thalamic cats was confronted by a fierce dog, the cat displayed the typical cat-like fear/rage response but this response switched-off the moment the dog was removed from sight. This rapid switching is not a normal characteristic of emotion in intact cats. In most animals emotion has an enduring quality.

The second major theory of emotion based upon structures in the central nervous system also has two main contributors and it is called the Papez-MacLean theory of emotion. It arose from Papez's (1937) insightful theoretical speculations that were based upon his clinical observations. He claimed that the limbic system was the structure in the brain concerned with emotion. Papez argued that, under cortical control, emotion built up in the hippocampal formation and transferred to the mammillary bodies, from these it travelled through the anterior thalamic nuclei to the cortex of the cingulate gyrus which was the receptive area for the experience of emotion (see Figure 3.10). Neural impulses, or radiations as he called them, flowed to other regions of the cerebral cortex adding emotional colouring. MacLean (1949) extended the theory and incorporated a number of clinical and research studies, arguing that the essential role of the limbic system was to add emotional colouring to experiences that enabled them to be interpreted in terms of emotional feelings rather than mere intellectualizations.

The third major theory that was very influential was Lindsley's (1951) arousal theory of emotion which was based upon the reticular activating system. Lindsley stated, and it is one of the weaknesses of his theory because every action becomes emotion, that there was a correlation between emotion and cortical arousal. Lindsley's arguments were based upon the fact that there was a correlation between emotion and cortical arousal but there is a similar correlation between behaviour *per se* and cortical arousal. He also argued that momentary block of EEG activity called alpha waves was also correlated with emotion. However this block might indicate a preparatory response. Finally, Lindsley's theory provided no means of determining different emotions.

If we consider these theories in chronological order, we discover that they reflect the increasing knowledge that was being gained over that period about the structure and function of the brain. The earlier theory of Cannon and Bard had four or five main neural pathways while later theories had something approaching ten times as many brain pathways. As investigators extended the areas of their search in the brain they discovered more and more areas containing emotional components. Apart from some attempts to give the hypothalamus a central place in mediating emotion this trend of increasing sophistication has continued.

Attempts to base a theory of emotion on the hypothalamus are interesting because they reversed the trend for increasing complexity and accorded one of the smallest areas of the brain a very large role in the mediation of emotion.

Perhaps it is not too surprising because for a brief period in the 1940s and early 1950s the hypothalamus was sometimes presented as masterminding some of the more interesting functions of the brain. This arose because of the discovery of the role of the hypothalamus in sex, hunger, thirst, temperature, and emotion. As we have discussed when we considered temperature control by the hypothalamus, the actual control by the hypothalamus is not as extensive as it first appeared. As has been argued earlier in the chapter, the confusion appears to concern confusion over the distinction between 'policy' and 'executive' areas of the brain. The group of theories and the arguments for and against them can be found in Alpers (1940), Gellhorn *et al.* (1940), Masserman (1941), and Wheatley (1944). This group of theories is best represented by Gellhorn who has carried out many experiments concerning the role of the hypothalamus and an integrated account is presented in Gellhorn and Loofbourrow (1963). Gellhorn's hypothalamic theory of emotion could be called the 'emostat' theory of emotion because it relies upon the concept of autonomic tuning proposed by Hess and Akert (1955). Gellhorn argues that emotion ranges along a continuum from excitement to depression and that this psychological continuum is controlled by a somatic/autonomic integration in the hypothalamus. Gellhorn's model is interesting in that it contains many feedback loops and an extensive relationship with the endocrine system which is ignored in most theories.

Gellhorn is able to explain how an extreme autonomic emotional state can flow over into the opposite state. He gives as an example the situation where a motorist brings his car to an emergency stop to avoid hitting a pedestrian. This reaction involves a state of maximum ergotropic or sympathetic alertness; however, the motorist is found only seconds later in a state of trophotropic or parasympathetic shock. Gellhorn argues that this comes about because the extreme emotional level in the sympathetic controlling areas of the hypothalamus 'overflows' into parasympathetic controlled areas of the hypothalamus. Despite his extensive and reasoned approach Gellhorn (1968) has failed to make any impact on theories of emotion. In fact, he fails to make the top twenty theories described by Strongman (1973). This is a pity for few theorists have built up such an imposing array of experiments and collected so much evidence. As Candland (1977) has indicated, it is difficult to predict what theories will survive; intellectual disposition or predisposition can be most fickle.

It seems that there is, at this time, no wish to revert to a central theory of emotion that is based largely upon a single area of the brain. Although in any future theory concerning the brain structure of emotion few can doubt that the hypothalamus will play a role. The essential problem of Gellhorn's theory seems to be that although he has assembled an impressive array of concepts from a number of unrelated fields it is not possible to justify incorporating them into a single model at the present time.

Recent theories of emotion have attempted to integrate emotion and cognition and the best example is the theory of Arnold (1970) who has proposed what she calls a phenomenological theory. Arnold's theory is notable because it attempts to integrate many aspects of emotion that have been ignored in the past three or four decades. Based upon physiological facts the theory is cognitive and conceives of an organism appraising an object or a stimulus. Emotion depends upon this action of appraisal which determines both the emotion and the action that is taken subsequently. While concerned with neural substrates of emotion Arnold has not discussed the autonomic nervous system explicitly but clearly it would play a role in the extensive brain interactions indicated by Arnold. The theory, while avoiding the simplistic mechanical view of many earlier theories, does separate emotion and reason but its inclusiveness makes it a rather difficult theory to test experimentally.

BIOSOCIAL THEORIES OF EMOTION

Schachter and Singer (1962) reported a series of experiments that produced a great impetus in the fields of social psychological research and provided a theoretical framework for a cognitive theory of emotion involving the autonomic nervous system. Schachter and Singer's experiments have been reported many times so it is not necessary to give more than the broad outlines of them. Their basic conclusion was that emotion resulted from an interaction between autonomic arousal and the perceived social environment.

The basic design of their experiments concerned the injection of adrenaline or a placebo substance into subjects who were divided into three groups. Subjects in one of the groups were injected with a placebo substance and these served as the control group. The other two groups were given injections of adrenaline and subjects in one of these groups were informed of the consequences of the injected substance while the other group of subjects were told that the injection was a vitamin substance that related to visual acuity. The groups were then placed into various social situations designed to arouse emotional reactions in the subjects. In certain of the experiments an experimental 'stooge' acted out prescribed emotional roles in the presence of the subjects. The conclusions from the experiments were that subjects who had received an injection of adrenaline, but told the true consequences of having such an injection, tended to respond with a greater degree of emotionality than the subjects who were not informed of the exact nature of the injection. The placebo group experienced little emotion. Schachter and Singer argued that when subjects are physiologically aroused but lack a ready explantation for their state they attribute it to the situation they find themselves in. The theory is able to explain how it is that a general state of autonomic activity can be expressed as different emotions. The theory agreed with the James-Lange theory but stipulated that autonomic arousal is a necessary but not always a sufficient producer of emotional experience. It explains how a general autonomic state can be turned into different emotions. This was one of Cannon's objections to the James-Lange theory. Schachter and Singer's theory provides a link between the apparently opposing views of James-Lange and Cannon-Bard.

Valins (1966, 1967a) suggested an important qualification to the Schachter and Singer model when he reported that subjects need not actually perceive their internal autonomic arousal in order to experience emotion. It was sufficient for subjects to think that an internal change had taken place. In other words, if a subject is presented with false information about his internal physiological reactions he behaves if there were a genuine alteration in his internal state. In Valin's experiments subjects were given false feedback information about their heart-rates while viewing slides of nude females. He reported that the perceived heart-rate, as opposed to the subject's true heart-rate, influenced judgements made by the subjects about the attractiveness of nude slides they were viewing. Valins (1967b) reported a follow up experiment in which, using a questionnaire, he preselected subjects as being emotional or nonemotional. His conclusion from this experiment was that subjects who had scored relatively high on the emotion questionnaire labelled the slides as more attractive or less attractive in terms of whether they had been given a false increase or a decrease in heart-rate. This is an important statement because it infers that subjects' cognition of their internal states is more important than the actual internal states themselves. Moreover, the reported findings suggest that it is possible to manipulate a subject's emotion by the simple expediency of using false feedback information.

Comments made earlier about problems associated with using single autonomic reactions as an indication of general arousal should make us a little wary and it would be important to have a most complete analysis of the heart-rate data. Unfortunately, as Harris and Katkin (1975) have reported, it is difficult from Valins's specifications of his measurements to decide what the heart-rates of his subjects were doing. Were the changes fundamentally overall accelerative or decelerative or were they biphasic showing an alteration in direction after an initial period? Goldstein *et al.* (1972) attempted to answer this question and monitored subjects' heart-rates in an attempt to determine if it was the cognition of the false heart-rate or the case that the false heart-rate caused the subject's heart-rate to change in the same direction as the false feedback. Goldstein and his colleagues reported considerable mimic effects showing that when subjects heard an increase in what they took to be their heart-rate they produced a corresponding increase in their actual heart-rate.

Harris and Katkin (1975) have reviewed a number of experiments that have attempted to dissect out the findings reported by Valins and they have concluded that the fundamental problem in all the studies is lack of precision in the concepts used. They give as examples words such as 'awareness', 'emotion', and 'autonomic arousal'. The investigations into Valins's reported phenomenon have revived the old confusions between emotional behaviour and emotional experience and have used the former, usually an autonomic response, to make a statement about the latter. Harris and Katkin also feel that another one of the main problems has been confusion about primary and secondary emotion. Primary emotion requires the function of the autonomic nervous system for the expression of emotion. Secondary emotion, or conditioned emotion as it has been called earlier in this chapter, is emotion that is assumed to have originally depended upon an autonomic reaction but to have lost this dependency by learning. Harris and Katkin maintain that the majority of these social studies have been concerned with secondary emotion and therefore it serves little purpose to attempt to use autonomic reactions as indices of emotional responses. Indeed, we may wonder about the conditioned reactions that could be invoked in certain subjects by using slides of nude males and females. Such variables can serve only to increase the variance in the data.

Mandler (1975) has put forward a theory of emotion that involves the autonomic nervous system and cognition. Mandler argues that the autonomic nervous system acts as a secondary system initiating evaluation of situations that require what he calls, a meaning analysis. The position taken by Mandler is that the autonomic nervous system constitutes the part of the nervous system that activates emotion. It is the autonomic nervous system that produces the general arousal or activity which, combined with the cognitive interpretive system, lends the special quality to certain experiences that are interpreted as emotion. The 'cold' emotion seen in acting is emotion in which the autonomic nervous system reaction is missing. One may wonder if the quality of empathy induced by an actor might not be related to the degree of autonomic activity he is producing. In a sense this says no more than the performer is approaching a very realistic performance when he is able to involve his internal autonomic system as well as his external muscle systems. Marañon (1924) spoke of his subjects experiencing 'cold' emotion. What Marañon had done was to elicit a part of the total internal activity but the situations used by Marañon lacked the social significance that was later provided in the Schachter and Singer experiments.

It is possible to think of situations in which autonomic and endocrine activity come to evoke emotional feelings that may occur in the absence of a discernible cognitive event, premenstrual or menstrual depression appears to be a good example. These are internally induced feelings that are often spoken of as deep and profound. One of the problems with Mandler's account that is similar to Lindsley's problem mentioned earlier is that all activation of the autonomic nervous system appears to be considered as emotion and there is no room for the concept of feeling. This is reflected in Mandler's index where there is only one reference to feelings as opposed to 19 for emotion. To give an example, postprandial torpor is activation of the large part of the autonomic nervous system; it is scarcely an emotion but it is a discernible feeling.

What are we to conclude from this chapter on biofeedback and emotion? The experimental literature at first sight tends to confuse the issues. In fact, in more pessimistic moments, one is inclined to agree with Artemus Ward, a nineteenth century American humorist, who wrote: 'The researches of many eminent scientific men have thrown so much darkness on the subject that if they continue their researches we shall soon know nothing.' During emotional states we certainly experience our autonomic nervous systems. But this is not the whole story, the field of emotion is complex and there are many different aspects that require detailed study and researchers beginning to think in terms of more comprehensive theories. This is true even of areas that traditionally have set rather narrow limits for their studies of emotion (Brady, 1971).

We can suppose that the more primitive biological functions of emotion were concerned with survival in a largely hostile environment and these primary emotions were mediated by the autonomic nervous system. But as the brain evolved, it did not leave emotion or the autonomic nervous system in the lower reaches of the brain but carried them into the evolving neural structures, each level adding new dimensions and broadening the original scope in subtle ways. If you wish to see emotion in full flower it is to man that you must turn. Man is the most emotional animal. This argument implies that emotion plays a role in the highest level of human intellectual achievement and, indeed, we can say that the hallmark of a work of genius is perfect synthesis between the intellectual and emotional processes. The intellectual process devoid of emotion is recognized as imitation, however clever (Van Toller, 1976). The emotion dimension throughout the brain is, as we have already argued, paralleled by a representation of the autonomic nervous system.

We might speculate that the essential difference between emotional behaviour and emotional experience reflects a difference in level of involvement by the brain. If I weep because a tiny piece of grit lodges in my eye, you will see the physiological response of lachrymation. If I weep because of an intense emotional experience, you will see the same physiological response of lachrymation but it has an additional important cognitive component. The two situations are quite different, the latter is mediated by the hypothalamus via the cerebral cortex; the former is mediated by the hypothalamus with little involvement by higher cognitive centres.

In the fields of both biofeedback and emotion we have in the past been too concerned to look for the structure of the phenomenon we are studying. Such an approach not only simplifies the experimental situation but also promised quick success. However, in most areas this was not to be and we are now coming to live with the complete problem. We have come to attempt to understand the functions as well as the structure. This has required understanding of the physiology as well as the psychological phenomenon and combining these with a biological overview. More recently we have to attempt to integrate cognitive and social factors into our concepts and models. Biofeedback is a good example of an area of recent concern where avoidable errors were made. It was not that the studies themselves were wrong or concerned faulty experimental design, it was just that they were irrelevant in terms of the total problem to which they claimed to be addressing themselves. Eventually biofeedback and emotion will contribute much to each other but at the present time few integrative studies have been attempted although each field deals with important aspects of autonomic function.

REFERENCES

Adolph, E.F. (1967). 'Ranges of heart-rates and their regulation at various ages (rat)', *Amer. J. Physiol.* **212**, 595-602.

Alpers, B.J. (1940). 'Personality and emotional disorders associated with hypothalamic lesions', in J.F. Fulton, S.W. Ranson and A.M. Frantz (Eds.) *The Hypothalamus and Central Levels of Autonomic Function.* Williams and Wilkins, Baltimore.

Arbit, J. (1957). 'Skeletal muscle effects of the chemical block of autonomic impulses and the extinction of fear', *J. comp. Physiol. Psychol.*, **50**, 144-5.

Arbit, J. (1958). 'Shock motivated serial discrimination learning and the chemical block of autonomic impulses', *J. comp. physiol. Psychol.*, **51**, 199-201.

Arnold, M. (1970). *Feelings and Emotions*. Academic, London.

Auld, F. (1951). 'The effects of tetraethylammonium chloride on a habit motivated by fear', *J. comp. physiol. Psychol.*, **44**, 565-74.

Averill, J.R. (1968). 'Grief: its nature and significance', *Psychol. Bull.*, **70**, 721-48.

Averill, J.R. (1969). 'Autonomic response patterns during sadness and mirth. *Psychophysiol.*, **5**, 399-414.

Barber, T.X., Kamiya, J. Miller, N.E., Shapiro, D., Stoyva, J. and DiCara, L.V. (1977). *1976-1977 Aldine Biofeedback and Self-control Annual*. Aldine-Atherton, Chicago.

Bard, P. (1934). 'On emotional expression after decortication with remarks on certain theoretical views', *Psychol. Rev.*, **41**, 309-329 (part I), 424-449 (part II).

Bard. P. and Mountcastle, V.B. (1948). 'Some forebrain mechanisms involved in the expression of rage with special reference to suppression of angry behaviour', *Research Publications of the Association of Nervous and Mental Disease*, **27**, 362-404.

Beatty, J. and Legewie, J. (1977). *Biofeedback and Behaviour*. Plenum Press, New York.

Birk, L., Crider, A., Shapiro, D. and Tursky, B. (1966). 'Operant electrodermal conditioning under partial curarization', *J. comp. physiol. Psychol.* **62**, 165-6.

Black, A.H. Operant conditioning of heart-rate under curare. *Technical Report No. 12, Dept. of Psychology, McMaster's University*, Canada (1967).

Blanchard, E.B., and Young, L.B. Self-control of cardiac functioning: a promise as yet unfulfilled. *Psychological Bulletin*, **79**, 145-163 (1973).

Bovet, D., Bovet-Nitti, F. and Marini-Bettolo, G.B. *Curare and Curare-Like Agents*. Elsevier, Amsterdam (1959).

Brady, J.V. (1953). 'Does tetraethylammonium chloride reduce fear? *J. comp. physiol. Psychol.*, **46**, 307-10.

Brady, J.V. (1971). 'Emotion revisited', *J. Psychiat. Res.*, **8**, 363-84.

Brener, J. (1977). 'Visceral perception', in J. Beatty and H. Legewie (Eds.) *Biofeedback and Behaviour*. Plenum Press, New York.

Brown, B.B. (1974). *New Mind, New Body. Biofeedback: New Directions for the Mind*. Harper and Row, New York.

Caldwell, D.F. (1962). 'Effects of adrenal demedullation on retention of a conditioned avoidance response in the mouse', *J. comp. physiol. Psychol*, **55**, 1079-81.

Candland, D.K. (1977). 'The Persistent problems of emotion', in Candland *et al. Emotion*. Monterey Brooks Cole Pub. Co., California.

Candland, D.K., Fell, J.P., Keen, E., Leshner, A.I., Plutchik, R. and Tarpy, R.M. (1977). *Emotion*, Monterey Brooks Cole Pub Co., California.

Cannon, W.B. (1929). 'The James-Lange theory of emotion', *Amer. J. Psychol.*, **39**, 106-24.

Cannon, W.B. (1931). 'Again the James-Lange and the thalamic theories of emotion', *Psychol. Rev.*, **38**, 281-95.

Cannon, W.B., Lewis, J.T. and Britton, S.W. (1927). 'The dispensability of the sympathetic division of the autonomic nervous system', *Boston Med. Surgical J.*, **197**, 514-5.

Cannon, W.B., Newton, H.F., Bright, E.M., Menkin, V. and Moore, R.M. (1929). 'Some aspects of the physiology of animals surviving complete exclusion of sympathetic nerve impulses', *Amer. J. Physiol.*, **89**, 84-107.

Carmona, A., Miller, N.E. and Demierre, T. (1974). 'Instrumental learning of gastric vascular tonicity responses', *Psychosomatic Medicine*, **36**, 156-63.

Church, R.M. (1964). 'Systematic effect of random error in the yoked control design', *Psychol. Bull.*, **62**, 122-31.

Dana, C.L. (1921). 'The anatomical seat of emotion: a discussion of the James-Lange theory', *Arch. Neurol. Psychiat.* (Chicago) **6**, 634-9.

Darwin, C. (1872). *The Expression of the Emotions in Man and Animals.* Murray, London.

Davitz, J.R. (1953). 'Decreased autonomic functioning and extinction of a conditioned emotional response', *J. comp. physiol. Psychol.*, **46**, 311-3.

Dawson, M.E. and Furedy, J.J. (1976). 'The role of awareness in human differential autonomic classical conditioning: the necessary-gate hypothesis', *Psychophysiol.* **13**, 50-3.

DiCara, L.V. (1970). 'Learning in the autonomic nervous system', *Scientific American,* **222**, 30-9.

DiCara, L.V. and Miller, N.E. (1968a). 'Instrumental learning of systolic blood pressure responses by curarized rats: dissociation of cardiac and vascular changes', *Psychosomatic Medicine*, **30**, 489-99.

DiCara, L.V. and Miller, N.E. (1968b). 'Long term retention of instrumentally learned heart-rate changes in the curarized rat', *Commun. Behav. Biol.*, Part A, 19-23.

Dunlap, K. (Ed.) (1967). *The Emotions.* Hafner, New York.

Eccles, R.M. and Libet, B. (1961) 'Origin and blockade of the synaptic response of curarized sympathetic ganglia', *J. Physiol.* (London). **157**, 484-503.

Falconer, W. (1796). *The Influence of the Passions Upon Disorders of the Body.* Dilly, London.

Favill, J. and White, P.D. (1917). 'Voluntary acceleration of the rate of the heart-rate', *Heart*, **6**, 175-186.

Fehr, F.S. and Stern, J.A. (1970). 'Peripheral physiological variables and emotion', *Psychol. Bull.*, **74**, 411-24.

Fuhrer, M.J. and Kilbey, M. (1967). 'Effects of spinal-cord transectomy on electrodermal activity in man', *Psychophysiol.* **4**, 176-86.

Furedy, J.J. (1971). 'Explicitly-unpaired and truly-random CS; controls in human classical differential autonomic conditioning', *Psychophysiol.*, **8**, 497-503.

Furedy, J.J. (1973). 'Some limits on the cognitive control of conditioned autonomic behaviour', *Psychophysiol.*, **10**, 108-111.

Garcia, J. and Ervin, F.R. (1968). 'Gustatory-visceral and telereceptor-cutaneous conditioning and adaptation in internal and external milieus. *Commun. Behav. Biol.,* part A, **1**, 89-415.

Garcia, J. and Rusiniak, K.W. (1977). 'Visceral feedback and the taste signal, in J. Beatty and H. Legewie (Eds.) *Biofeedback and Behaviour.* Plenum Press, New York.

Gellhorn, E. (Ed.) (1968). *Biological Foundations of Emotion*, Scott, Foresman, Illinois.

Gellhorn, E., Cortell, R. and Feldman, M. (1940). 'The anatomical basis of emotion', *Science*, **92**, 288-9.

Gellhorn, E. and Loufbourrow, G.N. (1963). *Emotions and Emotional Disorders.* Hoeber, New York.

Goldstein, D. Fink, D. and Mettee, D.R. (1972). 'Cognition of arousal and actual arousal as determinants of emotion', *J. Personality Social Psychol.*, **21**, 41-51.

Goldstein, M.L. (1968). 'Physiological theories of emotion: a critical historical review from the standpoint of behaviour theory', *Psychol. Bull.* **69**, 23-40.

Green, E. and Green, A. (1977). *Beyond Biofeedback.* 'A Merloyd Lawrence book' Robert Briggs and Associates, San Francisco.

Gruhzit, O.M., Fisken, R.A. and Cooper, B.J. (1948). 'Tetraethyl ammonium chloride: acute and chronic administration in experimental animals', *J. Pharmacol.*, **92**, 103-7.

Hahn, W.W. (1970). 'Apparatus and technique for work with the curarised rat', *Psychophysiol.*, **7**, 283-6.

Harris, A.H. and Brady, J.V. (1974). 'Animal learning: visceral and autonomic conditioning', *Ann. Rev. Psychol.*, **25**, 107-33.

Harris, V.A. and Katkin, E.S. (1975). 'Primary and secondary emotional feedback: an analysis of the role of autonomic feedback on affect, arousal, and attribution', *Psychol. Bull.*, **82**, 904-16.

Haygarth, J. (1800). *Of the Imagination as a Cause and as a Cure of the Disorders of the Body.* Cruttwell, Bath.

Herrnstein, R.J. (1969). 'Method and theory in the study of avoidance', *Psychol. Rev.*, **76**, 49-69.

Hess, W.R. and Akert, K. (1955). 'Experimental data on the role of the hypothalamus in mechanisms of emotional behaviour', *Amer. Med. Assoc. Arch. Neurol. Psychiat.*, **73**, 127-9.

Hestor, G.A. (1971). 'Effects of functional transection of the spinal cord on task performance under varied motivational tasks', *Psychophysiol.*, **8**, 451-61.

Hilgard, E.R. and Humphreys, L.G. (1938a). 'The retention of conditioned discrimination in man', *J. Exp. Psychol.*, **19**, 111-25.

Hilgard, E.R. and Humphreys, L.G. (1938b). 'The effect of supporting and antagonistic voluntary instructions on conditioned discrimination', *J. Exp. Psychol.*, **22**, 291-304.

Hohmann, G.W. (1966). 'Some effects of spinal cord lesions on experienced emotional feelings', *Psychophysiol.*, **3**, 143-56.

James, W. (1884). What is an emotion? *Mind*, **19**, 188-205.

Johnson, C.H. and Spalding, J.M.K. (1974). *Disorders of the Autonomic Nervous System.* Blackwell, Oxford.

Johnston, D. (1977). Biofeedback, verbal instruction and the motor skills analogy', in J. Beatty and J. Legewie (Eds.) *Biofeedback and Behaviour.* Plenum Press, New York.

Jonas, G. (1973). *Visceral Learning: Towards a Science of Self-Control.* Viking, New York.

Karlins, M. and Andrews, L.M. (1972). *Biofeedback: Turning on the Power of your Mind.* J.B. Lippincott Co., New York.

Katkin, E.S. and Murray, E.N. (1968) 'Instrumental conditioning of autonomically mediated behaviour', *Psychol. Bull.*, **70**, 52-68.

Kimmel, H.D. (1967). 'Instrumental conditioning of autonomically mediated behaviour', *Psychol. Bull.*, **67**, 337-45.

Kimmel, H.D. (1974). 'Instrumental conditioning of autonomically mediated responses in human beings', *Amer. Psychol.*, **29**, 325-35.

Lange, C.G. (1967). 'The emotions', in K. Dunlap, (Ed). *The Emotions.* Hafner, New York.

Levi, L. (Ed. (1975). *Emotions: Their Parameters and Measurement.* Raven Press, New York.

Levi-Montalcini, R. and Angeletti, P.U. (1966). 'Immunosympathectomy', *Pharmacol. Rev.*, **18**, 619-28.

Levine, S. and Soliday, S. (1962). 'An effect of adrenal demedullation on the acquisition of a conditioned avoidance response', *J. comp. physiol. Psychol.*, **55**, 214-5.

Lindsley, D.B. (1951) Emotion, in S.S. Stevens (Ed.) *Handbook of Experimental Psychology.* J. Wiley and Sons, New York.

Lipton, E.L., Steinschneider, A. and Richmond, J.B. (1965). 'The autonomic nervous system in early life. Parts 1 and 2,' *New England J. Med.*, **273**, 147-53.

Little, B. and Zahn, T.P. (1974). 'Changes in mood and autonomic functioning during the menstrual cycle', *Psychophysiol.*, **5**, 579-90.

McKilligott, J.W. (1959). *Autonomic Functions and Affective States in Spinal Cord Injury.* Unpublished Ph.D. thesis. University of California.

MacLean, P.D. (1949). 'Psychosomatic disease and the "visceral brain": recent developments bearing on the Papez theory of emotion', *Psychosomatic Medicine* **II**, 338-53.

Malmo, R.B. (1975) *On emotions, Needs, and Our Archaic Brain.* Holt, Rinehart, and Winston, New York.

Mandler, G., Mandler, J.M. and Uviller, E.T. (1958). 'Autonomic feedback: the perception of autonomic activity', *J. Abnormal and Social Psychol.* **56**, 367-73.

Mandler, G. (1975). *Mind and Emotion.* J. Wiley and Sons Ltd, New York.

Marañon, G. (1942) 'Contribution a l'étude de i'action emotive de d'adrenaline', *Revue Francais D'endocrinologie,* **2**, 301-325.

Masserman, J.H. (1941). 'Is the hypothalamus a centre of emotion?' *Psychosomatic Medicine,* **3**, 3-25.

Miller, N.E. (1967). 'Psychosomatic effects of specific types of training', in E. Tobach (Ed.) *Experimental Approaches to the Study of Emotional Behaviour. Ann. New York Academy Sciences,* **159**, 1025-40.

Miller, N.E. (1969). 'Learning of visceral and glandular responses', *Science,* **163**, 434-45.

Miller, N.E. (1973). 'Learning of glandular and visceral responses', in D. Shapiro (Ed.) *Biofeedback and Self-Control.* Aldine, Chicago.

Miller, N.E. and Carmona, A. (1967). 'Modification of a visceral response, salivation in thirsty dogs, by instrumental training with water reward', *J. comp. physiol. Psychol.,* **63**, 1-6.

Miller, N.E. and Banuazizi, A. (1968). 'Instrumental learning by curarised rats of a specific visceral response, intestinal or cardiac', *J. comp. physiol. Psychol.,* **65**, 1-7.

Mittelmann, B. and Walff, H.G. (1939). 'Affective states and skin temperature: experimental study of subjects with "cold hands" and Raynaud's syndrome', *Psychosomatic Medicine,* **1**, 271-92.

Mowrer, O.H. (1938). 'Preparatory set (expectancy) — A determinant in motivation and learning', *Psychol. Rev.,* **45**, 62-91.

Mowrer, O.H. (1947). 'On the dual nature of learning: a reinterpretation of "conditioning" and "problem solving",' *Harvard Educat. Rev.,* **17**, 102-48.

Mowrer, O.H. (1960). *Learning Theory and Behaviour.* Wiley, New York.

Mowrer, O.H. and Keehn, J.D. (1958). 'How are intertrial "avoidance" responses reinforced?' *Psychol. Rev.,* **65**, 209-21.

Moyer, K.E. and Bunnell, B.W. (1959). 'Effect of adrenal demedullation on an avoidance response in the rat', *J. comp. physiol. Psychol.,* **52**, 215-6.

Moyer, K.E. and Bunnell, B.W. (1960). 'Effect of adrenal demedullation operative stress and noise stress on emotional elimination', *J. Genet. Psychol.,* **96**, 375-82.

Murchison, C. (Ed.) (1928). *Feelings and Emotions; the Wittenburg Symposium.* Clark University Press, Mass.

Obrist, P.A. (1976). 'The cardiovascular-behaviour interaction as it appears today', *Psychophysiol.* **13**, 95-107.

Papez, J.W. (1937). 'A proposed mechanism of emotion', *Arch. Neurol. Psychiat.,* **38**, 725-43.

Pare, W.P. (1969). 'The effect of adrenalectomy, adrenal demedullation, and adrenalin on the aversive threshold in the rat', *Ann. New York Academy Sciences,* **159**, 869-79.

Plutchik, R. (1962). *The Emotions: Facts Theories and a New Model.* Random House, New York.

Pribram, K.H. and Luria, A.R. (1973). *Psychophysiology of the Frontal Lobes.* Academic Press, London.

Razran, G. (1972). 'Autonomic substructure and cognitive superstructure in behaviour theory and therapy: an East-West synthesis', in M. Hammer and K. Salzinger (Eds.) *Psychopathology.* J. Wiley, New York.

Rescorla, R.A. (1967). 'Pavlovian conditioning and its proper control procedures', *Psychol. Rev.,* **74**, 71-80.

Rescorla, R.A. and Solomon, R.L. (1967). 'Two-process learning theory: relationships between Pavlovian conditioning and instrumental learning', *Psychol. Rev.,* **74**, 151-82.

Rescorla, R.A. and Wagner, A.R. (1972). 'A theory of Pavlovian conditioning: Variations in the effectiveness of reinforcement and nonreinforcement', in A.H. Black and W. Prokasy (Eds.) *Classical Conditioning II: Current Research and Theory*. Appleton-Century-Crofts, New York.

Reymert, M.L. (1967). *Feelings and Emotions: The Moosehart Symposium* (Reprint 1967). Hafner, New York.

Schachter, S. and Singer, J. (1962). 'Cognitive, social, and physiological determinants emotional states', *Psychol. Rev.*, **69**, 378-99.

Schoenfeld, W.N. (1964). 'An experimental approach to anxiety, escape and avoidance behaviour', in P.H. Hoch and J. Zubin, (Eds.) *Anxiety*, Hafner, New York.

Schwartz, G.E. (1975). 'Biofeedback, self-regulation, and the patterning of physiological processes', *American Scientist*, **63**, 314-22.

Schwartz, G.E. and Beatty, J. (Eds.) (1977). *Biofeedback. Theory, and Research*. Academic Press, New York.

Shapiro, D. (1977). 'A monologue on biofeedback and psychophysiology', *Psychophysiol.* **14**, 213-27.

Sherrington, C.S. (1900). 'Experiments on the value of vascular and visceral factors for the genesis of emotion', *Proc. Royal Society*, **366**, 390-403.

Skinner, B.F. (1935). 'Two types of conditioned reflex and a pseudo-type', *J. Gen. Psychol.*, **12**, 66-77.

Skinner, B.F. (1938). *The Behaviour of Organisms: An Experimental Analysis*. Appleton-Century-Crofts.

Smith, K. (1954). 'Conditioning as an artefact', *Psychol. Rev.*, **61**, 217-25.

Smith, S.M., Brown, H.O., Toman, J.E.P. and Goodman, L.S. (1947). 'The lack of cerebral effects of d-tubcurarine', *Anaesthesiology.* **8**, 1-14.

Solomon, R.L. and Wynne, L.C. (1950). 'Avoidance conditioning in normal dogs and in dogs deprived of segments of autonomic functioning', *Amer. Psychol.*, **5**, 264.

Solomon, R.L. and Wynne, L.C. (1953). 'Traumatic avoidance learning: acquisition in normal dogs', *Psychol. Monogr.*, **67**, No. 4.

Spalding, J.M.K. (1965). 'The effects of psychological phenomena on autonomic function', *J. Psychosomatic Res.*, **9**, 149-53.

Stanley-Jones, D. (1966). 'The thermostatic theory of emotion: a study in kybernetics', *Progr. Biocybernetics*, **3**, 1-20.

Steiner, G. and Schonbaum, E. (Eds.) (1972). *Immunosympathectomy*. Elsevier, Amsterdam.

Stern, R.M. (1967). 'Operant conditioning of spontaneous G.S.R.: negative results', *J. Exp. Psychol.*, **75**, 128-30.

Sternbach, R.A. (1962). 'Assessing differential autonomic patterns in emotions', *J. Psychosomatic Res.*, **6**, 87-91.

Strongman, K.T. (1973). *The Psychology of Emotion*. Wiley, London. Second edition (1978).

Stunkard, A. and Koch, C. (1964). 'The interpretation of gastric motility: 1 Apparent bias in the reports of hunger by obese persons', *Arch. Gen. Psychiat.*, **11**, 74-82.

Sutherland, A. (1898). *Origin and Growth of the Moral Instinct*. In two volumes. Longmans, Green, London.

(1973). 'Symposium: Classical conditioning and the cognitive processes', *Psychophysiol.* **10**, 74-121.

Tarpy, R.M. (1975). *Basic Principles of Learning*. Scott, Foresman and Co., California.

Taylor, N.B. (1922). 'Voluntary acceleration of the heart', *Amer. J. Physiol.*, **61**, 385-98.

Thornton, E.W. (1971). *'Operant Heart-rate Conditioning in the Curarised Rat'*, Unpublished Ph.D. thesis. University of Durham.

Tobach, E. (Ed.) (1969). 'Experimental approaches to the study of emotional behaviour', *Ann. New York Academy Sciences,* **159**, 621-1126.

Trowill, J.A. (1967), 'Instrumental conditioning of the heart-rate in the curarised rat', *J. Comp. Physiol. Psychol.* **63**, 7-11.

Tuke, D.H. (1972). *Illustrations of the Influence of the Mind upon the Body in Health and Disease Designed to Elucidate the Action of the Imagination.* J. and A. Churchill, London.

Van Toller, C. (1976) *The function of human emotion.* Open Lecture, University of Warwick.

Van Toller, C. and Tarpy, R.M. (1972). 'Effect of cold stress on the performance of immunosympathectomized mice', *Physiol. Behav.* **8**, 1-3.

Van Toller, C. and Tarpy, R.M. (1974). 'Immunosympathectomy and avoidance behaviour', *Psychol. Bull.* **8**, 132-7.

Valins, S. (1966). 'Cognitive effects of false heart-rate feedback', *J. Personality Social Psychol.* **4**, 400-8.

Valins, S. (1967a). 'Emotionality and autonomic reactivity', *J. Experimental Research Personality.* **2**, 41-8.

Valins, S. (1967b) 'Emotionality and information concerning internal reactions', *J. Personality Social Psychol.* **6**, 458-63.

Watson, J.B. (1928). 'The heart or the intellect?' *Harpers Monthly Magazine,* **156**, 345-53.

Weiskrantz, L. (Ed.) (1968). 'Some traps and pontifications', in L. Weiskrantz (Ed.) *Behavioural Change.* Harper and Row, London.

Wenger, M.A. (1950). 'Emotion as visceral action: an extension of Lange's theory', in *Feelings and Emotions: the Moosehart Symposium.* McGraw Hill, New York. (Reprinted by Hefner 1967, see Reymert, M.L.).

Wenger, M.A., Bagchi, B.K. and Anand, B.K. (1961). 'Voluntary cardiac control', *Circulation,* **24**, 1319-25.

Wenzel, B.M. (1972). 'Immunosympathectomy and behaviour', in G. Steiner and E. Schonbaum (Eds.) *Immunosympathectomy.* Elsevier, Amsterdam.

Wheatley, M.D. (1944). 'The hypothalamus and affective behaviour in cats', *Arch. Neurol. Psychiat.* (Chicago), **52**, 296-316.

Wynne, L.C. and Solomon, R.L. (1955). 'Traumatic avoidance learning: acquisition and extinction in dogs deprived of normal peripheral autonomic function', *Gent. Psychol. Monogrs,* **53**, 241-84.

Zaimis, E. (Ed.) (1972). *Nerve Growth Factor and its Antiserum.* Athlone, London.

CHAPTER 6

Psychosomatic Disease: Genetic, Developmental, and Social Factors

In this final chapter we shall consider the relationship of the autonomic nervous system to psychosomatic disease. Clearly each of the different facets listed in the title of this chapter plays an integral role in both normal and abnormal functions of the autonomic system. Moreover, we find ourselves confronted by a dynamic whole because it is the interaction between these factors that makes the problem so intractable. However, by considering the parts we may come to gain insights that will aid our comprehension of the whole.

Prior to Freud, the medical profession, took its model of human physiology from the tremendous advances made in the physical and biological sciences during the nineteenth century and looked for explanations of mental illness in cellular pathology. Medicine conceived the body as a collection of physicochemical events in which the role of the doctor could be likened to the role of an engineer in that he became an expert in specific parts and systems of the body. Freud reintroduced the importance of mental aspects largely discarded since the Cartesian doctrine of duality. Gradually his ideas that mental or psychological events could affect the function of the body became prevalent. Emotion was the word used to describe this reaction and no better example of this type of conception can be found than in the massive annotated bibliography by Dunbar (1946) called *Emotion and Bodily Change*. However, the word emotion was used as a term having many meanings and it became apparent that its value was limited. One major confusion about the term emotion has always been that the word is used as a synonym for the word abnormal (i.e., for emotional disturbance you should read abnormal disturbance, meaning outside the range of normal responses as defined by social or cultural criteria). So the word emotion gradually dropped from favour and was replaced by the term psychosomatic to indicate an interaction between psychological and physiological aspects of the body.

Up to a period about the beginning of the 1960s the most popular method of studying psychosomatic complaints was to measure as wide a number of factors as possible (these might include socioeconomic, personality, etc.) and subject them to factor analysis with the object of arriving at a limited number of crucial factors (Alexander, 1950). More recently investigators have come to concentrate on specific physiological factors (Weiner, 1971). As pointed out by

Lader (1975) the main danger with the approach of eclecticism is that too broad a sweep will tend to dilute the chances of a genuine advance in the problem. The single aspect approach may result in attempts to locate the causation of a massive complaint in a discrete biochemical reaction. One major problem has been the lack of precision in the concepts used in psychosomatic medicine. Most psychosomatic research ignores the fundamental theoretical issue of the nature of the assumed interface between the parallel psychological and physiological processes and concentrates on one or the other aspect. Graham (1972) has stated that psychosomatic medicine is essentially clinical psychophysiology and indeed this is an important aspect but by no means a complete definition. What we find in psychosomatic disease is a discipline in which the two main approaches relate to psychological and physiological methods and have a basic interest in the structure and function of the autonomic nervous system.

Graham (1972) argues that it is wrong to think that most psychosomatic diseases have any special dependency on the pituitary-adrenal system. He also feels that statements about *physical* and *psychological* are statements about a choice to describe a process in one way or another. It is, he argues, always possible to describe psychological events in physical terms and, moreover, it is quite impossible to have a non physical state of an organism. The type of concept or theory used by an investigator is a matter of convenience or choice. Graham also raises problems about the impossibility of having a nonphysical state of the organism and these are illustrated by the various experimental techniques used to immobilize animals which are said to reduce physical factors. Immobilization techniques involve physical restraint and animals can usually be seen making efforts to escape. In addition, as indicated previously, when the drug curare is used to eliminate muscle movement we cannot be certain that the brain is not sending out signals that are producing effects in the peripheral parts of the body.

Contemporary theories relating to the autonomic nervous system and psychosomatic disease tend to fall into five main types of theories (Lachman, 1972):

1. *Personality type*: This type of theory takes a variety of forms but the basic argument in most is that the interplay or balance between the sympathetic and the parasympathetic nervous system is a major determinant of personality. Essentially it is a genetic theory.

2. *Constitutional vulnerability*: This is the weak link theory. In this type of theory the most vulnerable organ becomes the one to break down under stressful conditions. Of course, in certain individuals the weak organ is sometimes developed until it is the strongest. This concept of overcoming the weak organ was highlighted by the psychoanalyst Alfred Adler (1917) who pointed out that in certain persons considerable psychological compensation may take place for a weak physiological organ or structure. An example might be of a child with a weak heart who becomes a champion runner. This type of theory tends to focus on genetic vulnerability or early influences.

3. *Organ response learning:* this view holds that the major determining factor in physiological malfunction is that a learned reaction arises from an early

association between an emotional or traumatic event and a particular response pattern. This type of theory focuses on learning processes.

4. *Stimulus-situation theories*: basically this type of theory maintains that different emotional stimulus situations lead to different patterns of physiological response reactions that produce related damage in different organs of the body. Basically it supposes an innate relationship between various patterns of stimulation and various patterns of physiological reactivity in emotion. A good example is the classic study by Wolf and Wolff (1947) who directly observed gastric secretion of a patient with a gastric fistula. They claimed that fear caused a predominant sympathetic discharge and hostility and anxiety a predominant parasympathetic discharge. This type of theory has a genetic, learning process emphasis.

5. *Emotional reaction patterns*: this is a variation of the previous theory, the major difference being that emotional reaction theories tend to imply a cognitive component. This class of theory is sometimes called the individual-response pattern (Engel, 1960). Stimulus-situation theories provide a common model used in medicine in which a specific stimulus causes a certain response or sympton e.g., micro-organisms of the *Salmonella* type cause food poisoning. This type of theory has a genetic, learning process emphasis.

Having briefly introduced psychosomatic disease we can now turn our attention to the various factors mentioned in the title of this chapter in an attempt to gain some idea about their relative contribution towards autonomic structure and abnormal function.

GENETIC FACTORS

It is generally accepted that autonomic functions are influenced by genetic factors. A number of studies using identical twins have been held to confirm this (Jost and Sontag, 1944; Rachman, 1960). Jost and Sontag, during a three year study, obtained the following mean correlations for their groups: identical twins, +0.464; siblings, +0.316; unrelated group +0.087. From these results, they suggested the existence of an 'autonomic constitution' that had, at least, a partial hereditary factor. Levene *et al.* (1967) tested autonomic patterns of identical twins who were diagnosed as schizophrenic and failed to confirm the usual findings of similarity, but they attributed their results to the fact that the twins showed large behavioural differences. This later finding raises a query about the often quoted Jost and Sontag study, it is possible that the correlations they found in twins they used were due to similarities in their behaviour patterns. Wallace and Fehr (1970) compared Mongoloid children (Down's disease) with normal children, and found that Mongols showed fewer fluctuations in skin resistance, both under a normal and a distraction task condition. From their results it is not possible to determine if their findings were due to a secondary rather than primary autonomic disturbance. It is possible that Mongols were showing learned autonomic reactions.

The most specific and dramatic genetic complaint concerning autonomic malfunction is called dysautonomia (Vollmer, 1927; Riley, 1952). Transmission of familial dysautonomia has been shown to be caused by a recessive non sex-linked gene and both parents may be carriers. It has been estimated that the disease has a frequency of occurrence between 1/10,000 and 1/20,000 (McKusick *et al.* 1967). The disease is found in the children of Ashkenazim Jews (these people originally came from Northern and Eastern Europe. Sephardic and other oriental Jews are not thought to be affected). The list of symptons is horrendous and sufferers appear to have been brushed by most of the evils escaping from Pandora's box; patients are identified as suffering from a cluster of some of the known symptons. Virtually all patients suffer from lack of taste buds on their tongues and, although their eyes are normally moist, they do not produce additional lachrymation to emotional stimuli or mechanical irritation. This lack of tears is in sharp contrast to the excessive amounts of saliva and perspiration by these sufferers. Blood pressure and body temperature fluctuate wildly, tending overall to be high due to intense levels of vasoconstriction. Patients show poor coordination of breathing that is coupled to bouts of uncontrollable vomiting.

The psychological and behavioural symptoms shown by these patients are mainly but not wholly related to autonomic function. It is worth noting that dysautonomics show great difficulty in adapting to changes that involve diurnal rhythms. It is possible that this difficulty is related to the autonomic-pineal gland interaction but this possibility does not appear to have been investigated up to the present time. Dysautonomics have a narrow span of attention and they are unable to maintain sustained interest for more than very short periods. They tend to be hyperactive to impinging stimuli but paradoxically they are said to show reduced awareness of pain. It has been said that the reduced awareness of pain is not due to anaesthesia, they are able to locate pinpricks without difficulty, but due to an insensitivity to discomfort (Appenzeller, 1970). Dysautonomics show emotional liability to respond in an 'all or none' way. Coupled with their specific autonomic dysfunctions these patients show poor sensorimotor coordination with diminished tendon reflexes. Their speech is poor even though their hearing is said to fall within the normal range. Finally, as if these symptoms were not sufficient, dysautonomics may show mental retardation.

The cause of the disease is not known; but suggestions have ranged from abnormal development of autonomic neurones (Smith and Hui, 1973) to an enzymatic defect in either the cholinergic or noradrenergic systems (Riley, 1967; Gitlow *et al.,* 1970).

Because of its very clearcut nature it is hoped that this disease will provide some insights into the genetic foundations of autonomic functions. There must be strong possibilities of further categories of genetically linked psychosomatic complaints which remain still to be discovered. One interesting finding is a preliminary report of sex differences in the number of sympathetic neurones found in the spinal cord of cats (Calaresu and Henry, 1971). However, it will be some time before we will be in a position to decide if fundamental sexual differences exist in autonomic structures.

Schizophrenia is thought by a number of research workers to have a genetic component, Mednick (1970) and Mednick and Schulsinger (1968a, 1968b), as part of a longitudinal study on a group of children identified as having mothers suffering from schizophrenia, took psychophysiological recordings from them. They found, compared to the control group, that their experimental group had labile autonomic nervous systems responses. The experimental group were characterised by having GSRs with shorter latencies, increased amplitudes, and showing faster recovery times. Mednick argues that the labile autonomic systems of these subjects represent high levels of anxiety and this relates to his basic argument that schizophrenia is a learned response to overwhelming anxiety.

DEVELOPMENT OF THE AUTONOMIC NERVOUS SYSTEM

Earlier theorists believed that children when born were vagotonic, or parasympathetically dominated, and that sympathetic functions developed later. This belief appears to have arisen from the Victorians who believed that the first years of a child's life were spent eating, drinking, and sleeping. As mentioned previously, some early pharmacological evidence which was used to support this positiion (Eppinger and Hess, 1915) has now been shown to be incorrect. Richmond and Lustman (1955) reported qualitative and quantitative individual differences in autonomic function during the first few days of life. Lipton et al. (1965) have reviewed the evidence for developmental changes in two papers and they concluded that maturation changes in autonomic function takes place during late foetal and early neonatal life but the precise nature of these changes is masked by parallel maturational changes taking place in the central nervous system. Russian workers have reported that premature babies take longer to acquire responses that involve integration of autonomic and somatic components (Polikanina, 1961).

One limitation in most neonatal studies is that they are confined to the immediate postpartum period when the babies are still under hospital care. There appears to be a need for longitudinal studies in which the child's autonomic development could be monitored at various points. Lipton and his colleagues have concluded that significant changes in organ development occur during the early weeks and months following birth (Richmond et al., 1962; Steinschneider et al., 1964; Steinschneider and Lipton, 1965). If these changes were understood it would help us to understand how individual differences in autonomic function arose and give an indication of individual baseline levels. This type of longitudinal study produces problems relating to retest reliability, such as comparability of test conditions, but as mentioned earlier, Wenger (1966) has argued that it is possible to obtain reliable retest scores of autonomic function.

CONDITIONING AND CRITICAL PERIODS IN AUTONOMIC DEVELOPMENT

Certain psychoanalytic groups have argued the case for critical periods during the development of the autonomic nervous system being responsible for behaviour shown in later life. For example, Deutsch (1922) has proposed that physical symptoms shown by organs innevated by the autonomic nervous system are symbolic representations of suppressed emotions.

In contrast to the limited number of investigations using humans there are many accounts of experiments reported in the animal literature that show if neonates are handled patterns of behaviour are obtained in later adulthood that are different from patterns found in unhandled animals (King, 1958; Ader, 1959; Levine, 1962; Denenberg, 1962). In general, the majority of these studies have dealt with behavioural paradigms involving emotion and very few have looked specifically at the autonomic nervous system.

Two theories have been advanced to explain the handling phenomenon in terms of induced changes in the sympathetic nervous system. Bovard (1954) subjected young rats to immobilization stress and suggested that handling resulted in a reduction of the activity of the sympathetic nervous system accompanied by a permanent hypoactivity of the adrenal gland. In contrast to this view, Weininger (1954) reported that his handled animals showed evidence of increased sympathetic activity revealed by higher metabolic levels and greater distension of their cardiac vascular system. These diametrically opposing views at first sight appear puzzling but the differences could reflect peripheral versus internal differences in autonomic function, different handling procedures used, or a combination of both. Two workers have claimed that the critical variable in the handling phenomenon arises from loss of body heat when the young are removed from the mother (Hutchings, 1963; Schaefer, 1963). As discussed in Chapter 3, body temperature certainly relates to sympathetic activity and deserves a more detailed examination in future experimental studies involving neonatal handling.

At this point you may recall the Folkow and von Euler (1954) paper in which they suggested that the activity of the adrenal gland could be facilitated by electrical stimulation of the posterior hypothalamus, giving rise to the differential release of adrenaline and noradrenaline. This study was a limited and early report that requires replication but, again, it suggests, the possibility of altered sympathoadrenal activity via the hypothalamus that could arise from neonatal handling. This view receives some support from a study made by Levine (1966) in which dramatic changes were found in the sexual behaviour of rats following a single injection of the male hormone testosterone. Levine found that a single injection of the hormone produced permanent 'male-like' sexual behaviour in female rats which Levine argued was due to alterations produced in the brains of the animals.

AGEING

Very few textbooks concerned with ageing processes consider consequences arising from alterations in autonomic function and instead tend to concentrate on the more obvious and dramatic physical changes or pathologically revealed alterations in brain tissue. Yet it is possible that at least some of the problems associated with senescence are due to changes in autonomic function. Cannon

(1939, 1942) argued that in old people autonomic functions are held at normal values, but the total range is considerably reduced and old people become very susceptible to extremes at both the top and bottom ends of the range. For example, the body temperature of old people is maintained at 37°C but they are unable to cope with either hot or cold temperatures. This homeostatic alteration may be due to secondary effects such as changes in the skin but Cannon argued that the problem of adjustment was specifically related to basic alterations in the homeostatic process (Shock, 1952). Kuntz (1938) reported that even when secondary effects were allowed for, progressive structural changes could be shown in the ganglia cells of the autonomic nervous system that related to the ageing process. These alterations due to age included changes in both the cellular structure and content. Nelson and Gellhorn (1958) using the Mecholyl test, reported an increasing reduction in sympathetic and parasympathetic reactivity with age, but older people produce a relatively greater sympathetic reaction which, as a consequence, produces a sustained parasympathetic rebound. Thus, paradoxically it appears that older people are parasympathetically or vagotonically dominant.

Eisdorfer *et al.* (1970) presented a hypothesis that learning decrements found in older subjects is not simply due to changes in the central nervous system but is, in part, associated with the heightened activity of the autonomic nervous system that accompanies the learning tasks. Their hypothesis arose from the finding that if free fatty acids (Himms-Hagen, 1967) were taken as an indication of autonomic arousal, there appeared to be a curvilinear relationship between learning and autonomic activity. The effect of increased autonomic activity was to interfere with the learning task. Eisdorfer and his colleagues used the drug propranolol to lower autonomic activity. As indicated in Chapter 4, the action of this drug is partially to block the (nor)adrenergic receptor sites. They found that their experimental group (mean age 68.8 years) showed a significant decrease in number of errors made. This is an interesting finding that might be held to be supported by the earlier finding by Nelson and Gellhorn of increased sympathetic reactions in older subjects. Nelson and Gellhorn (1958) extended and confirmed their finding by using a group of psychotic patients.

Clearly further research is needed to discover the exact nature of the changes in autonomic function that could be producing alterations in learning performance. One possible overlooked variable in the Eisdorfer study was that propranolol, in common with many other drugs, may have caused hypothermia in their experimental subjects and it was a fall in body temperature that produced the superior performance. Within certain limits, temperature has been shown to influence performance in learning tasks.

SOCIAL, CLIMATIC, AND CULTURAL FACTORS

This section concerns areas where it is difficult to state with any precision whether we are observing genetic factors or learned patterns of social response that are used unconsciously by the population for their adaptive value.

An example of a physiological change in autonomic function was discovered when attempts were made to settle pygmies outside their normal habitat in the humid rain forests in Central Africa. The failure was said to be due to the fact that pygmies have a smaller number of sweat glands per unit of skin and this prevented them from dissipating sufficient body temperature in hot, low humidity conditions in which they were to be settled. If these reports are true, then it appears that the pygmy race has either evolved a special physiological mechanism, or undergone a reduction in the function of sweat glands to adapt the race for living in hot and humid conditions. Page and Brown (1953) found that compared to Caucasians, Eskimos had a greater flow of blood in their hands and forearms over a wider temperature range. As pointed out previously the normal physiological reaction to cold is vasoconstriction of blood vessels in the skin to reduce heat loss from the body and protect the vital inner core temperature. Prolonged and severe vasoconstriction of peripheral limbs can result in gangrene, so in a race that is constantly exposed to low temperature we might expect to find modification in the sympathetic nervous system to reduce effects arising from prolonged vasoconstriction of blood vessels. In the Page and Brown study, we cannot be sure if the phenomenon is due to a genetic factor or a learned response, because it is possible that Eskimos maintain a certain level of vasodilation in their limbs by behavioural manipulation.

The next example illustrates more obviously the possibility of a social habit that results in either the actual lowering of autonomic activity or the belief that it does, in order to carry out heavy work in extremely adverse conditions. It concerns Peruvian Indians who work in high-altitude silver mines in the South American Andes. It is the practice of these workers to chew leaves of the plant *Erythroxylon coca* which contains cocaine. The habit has ancient origins and was noted by Pizarro's expedition in the sixteenth century. It is believed by the Indians to help them undertake their arduous work at high altitudes (Burn, 1971). During the last century, investigators of this phenomenon held that the action of the ingested cocaine (see Figure 4.4, for its action in limiting uptake of noradrenaline on the receptor side of sympathetic nerves) was to stimulate the cortex and to elevate mood, temperature, and metabolism giving increased endurance against cold and fatigue in the thin atmosphere of the silver mines. However, in contrast to drug addicts who take cocaine, the habit of chewing the leaves is not addictive for it is discontinued if the Indians migrate to lower altitudes. Also, the amount of cocaine obtained by chewing the leaves is very small and oral ingestion is not an efficient way of getting the drug into the circulation. Hanna (1970) examined the physiological responses of coca chewing Indians with those who did not chew the leaves was unable to find any evidence of superiority of the former over the latter. Hanna felt that it was possible that if his findings were projected into maximally adverse conditions that some enhancement might be obtained from the small levels of cocaine ingested. So, the suggestion remains that the effects of chewing the leaves owe more to a social belief than a genuine physiological effect.

The findings reported in the last section might lead us to expect to find ethnic or cultural differences in autonomic function and evidence has been found to support such a view. Tursky and Sternbach (1967) and Sternbach (1966) have reported marked differences between different ethnic groups in terms of response sets towards pain: Yankees were said to display a phlegmatic attitude; Jews expressed concern for the implications of the shock and distrust of palliatives; Italians expressed a desire for pain relief; while the Irish inhibited expression of pain and suffering. The authors suggested a genetic or learned cultural pattern response set. However, the authors found large intragroup variability that prevented the means of the different groups from reaching significant levels.

CLIMATIC FACTORS

Other factors that influence autonomic activity are not so easy to classify. For example, Wenger (1966) has reported that he was unable to obtain reliable retest measures of autonomic function if the measurements were made during the mild southern Californian winter period. Those of us who spend our lives in more northern climes can testify to the reduced physiological functions, that are not solely due to low temperatures, during the long dark winter months. This depression of bodily function is presumably related to the pineal-autonomic interaction. An example involving heat comes from southern Germany where at certain times of the year, they suffer from what is called the föhn problem (Stangl, 1952). The föhn is a hot wind that blows down from the Swiss Alps and in its wake brings increased suicide rates and widespread psychological depression that are said to be related to changes in autonomic function.

VAGARIES IN THE STRUCTURE AND FUNCTION OF THE AUTONOMIC NERVOUS SYSTEM

In various sections of this book we have hinted at the structural and functional aspects of the autonomic nervous system that contribute towards it having what might be called a 'high idiosyncratic rating'. Not only do we find marked variabilities, but we also find the system has a remarkable ability to compensate for imposed reduction of its functions. As mentioned in Chapter 1 Romer suggested that the instability of the autonomic nervous system arose from an imperfect matching together of the peripheral and the central nervous systems at the evolutionary point where the central nervous system evolved as an entity. This line of reasoning suggests that psychosomatic disease may be a heritage of an evolutionary error. Did the brain develop in a way that prevented it from being truly compatible with the autonomic nervous system? These are questions to which we cannot give a definite answer at the present time and perhaps will never be able to. Nevertheless, we would be wise to take account of these possibilities when we are considering the influence of the autonomic nervous system on psychosomatic complaints. Wolf (1963) has suggested that psychosomatic disease represents too much or too little of the normal and essential adaptive reactions of the body. Alvarez (1940, 1951) pointed out that there are probably many individuals who have been classified as suffering from psychosomatic ailments when in fact they were suffering from genuine abnormal autonomic function. Had these abnormal functions been diagnosed as such, they would have, at the least, allayed the attendant fears of the sufferer and perhaps have resulted in alleviation of the actual condition. As individuals we all tend to consider ourselves to be essentially 'normal' and look askance at anyone who appears to be different from us. This type of problem will only be fully appreciated when longitudinal studies of autonomic function are made on individuals. As suggested earlier, ideally this type of screening could routinely be carried out at various periods during the life of individuals.

The vast majority of biofeedback studies concern attempts to reduce threshold levels of autonomic functions but there is most probably a class of patients who require the threshold of their autonomic activity to be raised. For this type of patient the problem arises from too little rather than too much autonomic activity. For example, can we train a person suffering from hypotension to raise his blood pressure? As mentioned in the last Chapter, Schwartz (1975) has indicated that it may be possible to train patients to raise their blood pressure.

Varni *et al.* (1971) discovered a little appreciated problem when they measured pethysmographic changes from either hands of 16 subjects and found bilateral differences that were not related to handedness. They suggested that their findings indicate that asymmetry of autonomic activity is typical rather than atypical for normal individuals. This interesting finding complements many others about the vagaries of the autonomic nervous system and must certainly modify the conclusions from certain research findings that had overlooked possible bilateral differences in subjects.

The major conclusion to draw from this section is that any investigator looking at behaviour and autonomic activity must be prepared to find large variations and contradictions in his data. Failure to appreciate this may mean a carefully planned experiment is invalidated.

PERSONALITY AND THE AUTONOMIC NERVOUS SYSTEM

The experimental evidence for the concept linking personality to the function of the autonomic nervous system was collected by Eppinger and Hess (1915). They advanced the idea, now enshrined in popular opinion that a person with a dominant sympathetic nervous system (sympathotonic) would display emotionally unstable characteristics; being impetuous, tense and restless, and socially dependent on others. Conversely, a person with a dominant parasympathetic nervous system (vagotonic) showed emotional stability; being deliberate and phlegmatic in action and socially independent. This viewpoint is represented by Kempf (1920), Kuntz (1951), Servais and Hubin (1965), and in a more sophisticated psychological form by Eysenck (1953). From the account given in Chapter 4 you will be aware of Wenger's measurements of autonomic balance. By taking autonomic measurments from USAF recruits, Wenger (1948) was able to show in terms of sympathetic or parasympathetic dominance that a man's autonomic bias would result in certain types of accidents. Men with a basic resting pattern of parasympathetic dominance tended to become involved in aircraft accidents because they failed to notice critical dials or take important readings. A sympathetic dominant personality tended to be involved in aircraft accidents because he over-reacted to a critical situation. Eysenck (1968) has reviewed the evidence for the autonomic nervous system determining the type of personality exhibited by a person and has concluded, 'the connection is not as firmly based in all its aspects as might at first appear'.

Why is this and what are the special problems that research workers in this area encounter? Workers looking for correlations between the autonomic nervous system and personality have tended to define one or other autonomic function as a behavioural response and then attempted to establish a correlation between the defined response and a score on a personality scale. Having done this most investigators then succumb to the undoubted temptation to generalize their findings to broad categories of behaviour. If we examine the particular autonomic function being measured we usually find that the measurement is confined to a single aspect of that function. Had another aspect of autonomic function been selected, it is possible that the converse conclusions might have had to be drawn. All too frequently autonomic functions are not examined in sufficient depth or breadth. A second major problem in this field concerns the various concepts of personality. A research worker in this field (Roessler, 1973) has pointed out that there are serious problems concerning the quantification, validation, and reliability. To these we must add the intrinsic and intractable problem of manipulating a variable like personality.

Over the years theories relating personality to the autonomic function have become more complex and recent explanations have tended to be grouped around nebulous physiological concepts like arousal (Eysenck, 1968). More recently it has been suggested that the concept of arousal be abandoned for 'conceptual nervous systems' (Gray, 1971). One danger with 'conceptual nervous systems' is, unlike the concept of arousal that can be shown not to have a firm physiological foundation, we encounter problems of definition and may find ourselves in the same situation as Alice, when she attempted to hold a conversation with Humpty Dumpty. It means what the author says it means, and no more; each author has a personal set of definitions from which it is difficult to generalize.

Rather than take a global concept like personality, Roessler (1973) has reported a series of experiments in which he attempted to analyse the effect of the autonomic nervous system on performance, by testing responses that related to its role as an essential homeostatic mechanism in the body. He did this by using a single personality factor that he feels relates to adaptation. The factor he took was reality testing as measured from the Minnesota Multiphasic Personality Inventory. It is defined as the ability to appraise accurately the nature and intensity of stimuli. Roessler argues that subjects with low ego-strength achive homeostasis by greater intensity of their activating and deactivating forces within the body. To quote his example; when a low ego-strength and a high ego-strength subject are brought into the laboratory to face a novel and strange situation it is often found that they show similar levels of autonomic activity. These comparable levels are achieved by using different amounts of energy. An analogy would be to see a heavy suitcase carried by a big man and a small man. Behaviourally they might both carry the case at the same speed, but the large man with his greater strength would be using less energy. A high ego-strength subject, having assessed the situation, will feel completely at ease and show low levels of activity that reflect his calm state. The low ego-strength subject feels anything but happy and is using a lot of energy to hold down his autonomic activity to preserve his defence mechanisms. This idea has been put forward at a number of points in this book and relates to whether a person showing predominant sympathetic nervous activity is, in fact, parasympathetic dominant but this is hidden by overlaying sympathetic activity that is brought about to repress the parasympathetic system. An explanation, outlined earlier in this chapter was advanced by Nelson and Gellhorn (1958) to explain the apparent parasympathetic dominance in old people. As Roessler points out, this type of factor would serve as a confounding variable in the experimental situation used by Lacey to explain heart-rate speeding up and slowing down as essentially an attentional mechanism.

Stewart and Dean (1965), using the unstructured personality test known as the Rorschach inkblot test, measured a number of autonomic responses and found a series of intercorrelations. These were said to be related to the way an individual was receptive to his environment and also to the way an individual categorizes or interprets sensory data. However, the authors also found some inconsistencies between the traditional interpretation of Rorschach scores and autonomic nervous system responses of their subjects.

Apart from personality variables, workers have looked at the influence of the autonomic nervous system on perception. In Chapter 3 the reflex role of the sympathetic nervous system in protecting the eye from excessive light was mentioned. Hess (1972) has used pupillary size of the eye as an index of the interest a subject is showing in what he is looking at. For example, if illumination is held constant and a subject is shown a picture he finds interesting or attractive, then his pupils will dilate. Hess has also claimed that if two practically identical pictures of a pretty girl are shown but in one picture the pupils are slightly enlarged, this will be judged as the most attractive picture by the majority of raters.

In an attempt to directly test functions of the sympathetic nervous system on perception, Callaway and Thompson (1953) stimulated the sympathetic system by using amylnitrite inhalations and the cold pressor test. They found that when these procedures were administered, subjects showed a consistent decrease in judging the apparent size of a distant object that, it was claimed, was not due to local opthalamic effects or decreased attention. They claimed that their finding related to narrowing of awareness resulting in reduction of the subject's perception of distance cues. The authors contrast their findings with Eppinger and Hess who suggested positive feedback in the autonomic nervous system by claiming to show a negative feedback function in their results. Interestingly, again, we find this concept of an inverse function in autonomic activity. The basic idea is that the apparent dominant system is masking a true dominant activity of the alternative system.

Another important point has been made by Mandler *et al.* (1958) who pointed out that in terms of reports of high autonomic activity a high scorer may represent a person who actually exhibits greater activity or a person who is more sensitive to autonomic activity. They found that high scorers showed significantly greater autonomic activity than low scorers; moreover, they also tended to overestimate their autonomic responses.

In recent years there has been an increase in the number of studies that have attempted to examine the social aspects of psychosomatic disease (Levi, 1971; Heine, 1972). These studies relate to social stress and human ecology and many of the papers reported are concerned to investigate individual reactions to this type of stress. Current conceptions tend to the view that genetic constitution determines how an individual will react but that the environment determines the stress to a large degree. One important development is the use of life style questionnaires (Rahe, 1975) that deal with reports of family crises, economic changes, social and work changes, and illnesses etc. Changes in life style have usually been taken to mean psychological changes but this need not be the case and changes may involve physical factors such as alterations of diet or different mineral contents of the water that might be experienced on a holiday. One of the most complete studies of using this type of research concerns an investigation made on an American warship (Rubin *et al.*, 1972), in which records were kept of the men on board during a lengthy voyage. The naval personnel were required to fill in a detailed questionnaire and this information was later used to check against the frequency that the men reported sick during the voyage. It was found that the incidence of sickness correlated highly with changes in reported life style over the preceeding six months.

There are clearly problems in this type of research in which assessments are made retrospectively and Sarason *et al.* (1975) have pointed out various weaknesses concerning the questionnaires in terms of psychometric validity and pointed out that the correlations produced in the earlier life style questionnaire research are low. This area of research into the complex ecology of human life is certain to increase in the future but we must not lose sight of the fact, certainly in terms of autonomic function, that the reactions to life style change are mostly natural bodily reactions and we must be careful about labelling them 'abnormal'. The problem breaks down into two parts, the first concerns responses that are exaggerated or insufficient and the second is the separate question of human ecology and what we can do about making it more conducive in terms of human needs.

AN AUTONOMIC NERVOUS SYSTEM LEARNING THEORY OF PSYCHOSOMATIC DISEASE

Both Sternbach (1966) and Lachman (1972) have given accounts of psychosomatic disease involving the autonomic nervous system and using behaviouristic models. Indeed, Sternbach has stated that an operational definition of psychosomatic disease is pathological function of the autonomic nervous system. Lachman has stated that with few exceptions, psychosomatic complaints are learned. It is often assumed, either explicitly or implicitly, that autonomic reactions cannot be learned consciously. However, it has been a constant theme throughout this book that autonomic responses can be learned. The essential point is that somatic and autonomic response patterns are intermingled with each other. As we have seen in the section dealing with biofeedback, it is possible to learn to regulate autonomic functions. One reason in the past for the relative lack of interest in autonomic functions as learning phenomena appears to be that autonomic functions are not readily observable in a freely moving animal performing a learning task.

Lachman, without minimizing the role of genetic factors, states that the essence of his theory is that psychosomatic manifestations are the result of frequent, or prolonged, or intense implicit reactions elicited via stimulation of receptors.

The sufficient conditions for the development of psychosomatic complaints have been put into a logical IF-THEN proposition by Sternbach (1966):

$$\text{If} \left\{ \begin{array}{l} \text{Individual} \\ \text{Response} \\ \text{Stereotypy} \end{array} \right\} \text{And} \left\{ \begin{array}{l} \text{Inadequate} \\ \text{Homeostatic} \\ \text{Restraints} \end{array} \right\} \text{And} \left\{ \begin{array}{l} \text{Exposure to} \\ \text{Activating} \\ \text{Situations} \end{array} \right\} \text{Then} \left\{ \begin{array}{l} \text{Psychoso-} \\ \text{matic} \\ \text{Episodes} \end{array} \right\}$$

This suggests that if an individual with a marked response stereotypy (may be genetic and / or learned) becomes involved in situations that regularly involved an autonomic response and the homeostatic threshold of this response is not sufficient to cope with the imposed stress, then it is likely that psychosomatic episodes will arise. This model can be used to try to understand certain aspects of psychosomatic disease. It does not answer the problem of which precedes which and we are left with the classic 'chicken and egg' puzzle. Moreover, this type of model is mainly heuristic allowing us to identify the sufficient factors but will probably not enable us to discover causal factors. Clearly learning is an important aspect but no more so than the genetic and the environmental aspects. In psychosomatic disease we are confronted with a complex ecosystem that contains genetic and learned components of the behaviour and we must produce models that enable us to comprehend the many and subtle interactions.

THE RELATIONSHIP BETWEEN THE SYMPATHOADRENAL AND THE PITUITARY-ADRENAL SYSTEMS

In Chapter 2 we looked at a possible biochemical relationship between the sympathoadrenal and the pituitary-adrenal systems pointing out that the concept of Cannon concerning the homeostatic role of the sympathoadrenal appeared to have common features with the general adaptation syndrome concept of Selye. Selye himself has been quoted (Mason, 1971) as saying that Cannon was never convinced by the importance of the pituitary-adrenal system. As Mason points out we need some link between these two important bodily systems. What might this link be?

Selye's (1956) concept was that the pituitary-adrenal system, being a nonspecific endocrine reaction, reacted in a general way to a diverse variety of stressors. It was thought initially to be a purely physiological reaction. However, it was shown that the pituitary-adrenal system reacted to stress situations in which great care had been taken to eliminate physical factors. For example, animals subjected to immobilization stress in which care had been taken to reduce physical factors showed a general adaptation syndrome of increased adrenal cortical activity. Mason concluded from a number of studies originally designed to reveal the effects of physical factors, that if he eliminated all the psychological factors surrounding a stressor then the general adaptation syndrome was not found. In support of his statement he cites an experiment designed to show that the physical effects of fasting evoked a general adaptation reaction. He found that if experimental monkeys were removed from their home cage, where they were unable to see other animals eating, and they were given a nonnutritive substance to eat they did not show a general adaptation reaction. Mason argues that his results suggest that in this area of research, attention should be directed away from the original emphasis of many diverse stimuli eliciting a 'nonspecific' endocrine response, towards the idea of many diverse stimuli eliciting a 'nonspecific' behavioural response. He asks, 'What bodily mechanisms are known to exist which might convey a quality like "noxiousness" to neuroendocrine centres?' His proposal is that on top of the normal homeostatic mechanisms of the body reacting in relatively specific patterns, there comes a point where the organism perceives a threatening physical situation of more than usual intensity. At this point the body brings universal endocrine responses into play. An overresponse to psychological stimuli may also come about because in such a situation the body is not sure what is the correct level of response that will be required. Reaction to a physical stimulus produces a graded response by the body which is able to 'know' what is the correct level of reaction. A psychological stimulus, particularly if it has not been encountered before, is more likely to produce an overreaction because the body will produce a maximum response in order to ensure that its response is adequate.

We might wonder about the possible roles of the pituitary-adrenal and sympathoadrenal systems at the behavioural level. In terms of adaptive behaviour an animal that is learning to successfully escape and avoid a threatening situation has to: (1) make an initial escape response; and (2) avoid the source of the danger. It is open to speculation that the pituitary-adrenal and the sympathoadrenal systems are the two mechanisms in the body that facilitate this behaviour. The faster reacting sympathetic system, with its nervous network, may act as the mechanism for making the initial *escape response* while the slower reacting pituitary-adrenal system (Fortier *et al.,* 1959) serves to maintain the essential flat generalization gradients of *avoidance responding*? A physiological process subserving avoidance responses needs to be active over relatively long periods of time. The slower acting pituitary-adrenal system would be well fitted for such a role.

Graham (1972) argues, 'It is a mistake to think that the essence of psychosomatic medicine is to show that Psychologically described states cause Physically described states.' But is this always the case? Animal experiments are now being reported in which attempts are being made to tackle the physical/psychological problem. Weiss (1971) subjected rats to repeated electric shocks using the 'yoked' design explained in Chapter 5. In his experiment electric shocks were proceeded by either (1) a warning signal, (2) a series of signals having a 'clock' function, or (3) no signal at all. Weiss found that a discrete warning signal reduced the level of ulceration both in animals having control over the shock and the helpless 'yoked' animals. He proposed a theory to explain his findings that relates to the level of coping responses made by the animals to the shocks. He predicted increased ulceration levels in animals making coping responses with low levels of feedback. That is, the critical factor causing ulcers in his conflict situation was where an animal made high level coping responses with low levels of feedback. If this is so then it appears that it is more beneficial, in terms of ulcers, for an animal in the experimental situation used by Weiss to passively accept the shocks without attempting to cope with them.

Weiss had produced a fascinating study and reading it one is reminded of the statement made by Ramey and Goldstein (1957), that the joint surgical operation of sympathectomy and adrenalectomy results in an animal that (within the sheltered confines of the laboratory, at least) is better able to withstand stress than an animal in which only one of the systems has been removed. This physiological finding appears to complement the point made by Weiss, by showing removal of both of these important physiological systems serves to eliminate 'physiological coping reactions' that may in the long run be more detrimental to the animal than the stress itself. This situation arises, as mentioned in connection with Mason, because in the doubly operated animal the adaptive demands on the body decrease, and energy is not being used to maintain states that are not absolutely vital. One way to survive is to become a passive recipient of the stress. This view has also been put forward by Bettelheim (1959) in his account of inmates of German concentration camps. Lazarus (1966) has also put forward a cognitive-phenomenological explanation of psychological stress and human behaviour that concerns the type of coping responses made by an individual. Seligman (1975) is concerned with helplessness and coping responses and he introduces an interesting approach from a behaviouristic point of view.

THE STIGMATISTS

One of the best examples demonstrating the complexity of psychosomatic disease and revealing our basic lack of understanding is the phenomenon of stigmatists who are people who are able to reproduce the wounds of Christ on their bodies. Over the centuries there have been more than 300 recorded cases of stigmatists and, given human behaviour, we may suppose that some of the cases concern people who deliberately inflicted the wounds on themselves. However, the fact remains that there are a sizable number of these cases that have been subjected to careful study in which self-inflicted wounds do not appear to be possible; that is, if we are not prepared to think that a large number of people have colluded together in an attempt to deliberately deceive. Stigmatists, who may be either male or female, are generally reported as being very devout, extremely sensitive, and very emotional people. The various investigations into cases of stigmatism have usually concluded that the wounds are the result of remarkable control of mind over body; the mind wills the body to reproduce the wounds of Christ and it does.

The most famous stigmatist of recent times was Padre Pio who was the first recorded ordained priest to bear all five marks of the stigmata. Padre Pio was born into a large peasant family and as the result of what was said to be a devout nature he joined the Franciscan order of friars to study for the priesthood when he was a teenager. There are no records of any physical abnormality until the morning of 20 September 1918 when he fainted during prayers and, on recovering consciousness, found that his hands and feet were bleeding. Later he found that he was also bleeding from a wound in his side. Brother friars found that they were unable to stop the bleeding. It was later said that he lost sufficient fluid from his side each day to saturate five handkerchiefs. After an extensive period of investigation by medical and ecclesiastical authorities it was concluded that the happy-go-lucky, rotund Padre Pio had produced the wounds of Christ.

What are we to make of cases like this? We may decide that science cannot sensibly comment on such cases but they remain to mock our ignorance and puzzle us. We may label the phenomenon as hysterical reactions but this procedure of category labelling does not solve the basic puzzle. For one thing it is difficult to understand how these open wounds remained free from infection long before the discovery of antibiotics. The reported absence of wounds on Padre Pio's body after his death in 1968 adds to the mystery. Stigmatists remain a mystery still to be explained and, at the very least, serve to remind us of the complexity of the body/mind relationships.

Henry and Ely (1976) have produced a diagram relating the processes of the brain to psychosocial stimulation and this diagram is reproduced in Figure 6.1. The figure relates to the flow of information in the brain and its integration at the three levels conceived by MacLean. The limbic system is given an important role in their model integrating emotion. Figure 6.1 relates to our neuroanatomical knowledge of the brain but it must be remembered that additional information is now available concerning chemical pathways of the

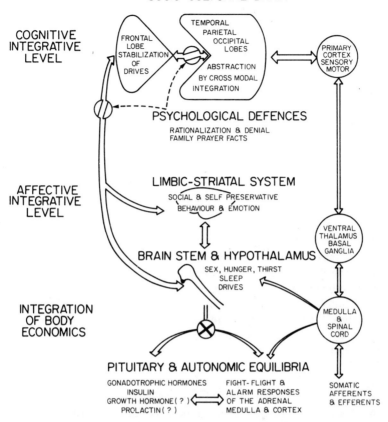

Figure 6.1 Suggested model for the 'socio-cultural' brain. Proposed physiological structures involved in psychosocial responses. After Henry and Ely. (1976), in L. Levi (Ed.) *Emotions: Their Parameters and Measurements*, Raven Press, New York.

brain. Watts (1975) has claimed that a large number of medical disorders involve chemical malfunction in specific parts of the brain and that the interaction of the mood and autonomic system occurs in the limbic system. It is the latter that is responsible for the symptoms of visceral illnesses. Watts suggests the name multisystem diseases as being more appropriate than the label psychosomatic. The figure tends to suggest three distinct levels but it must be remembered that the drawing is heuristic and that the brain and the autonomic nervous system are interdependent. Our genetic constitution, our physiology, our learned reactions are essentially part of us and all of these aspects are revealed in our responses. In addition, human interdependence also extends to environment of the individual and the diagram does little more than hint at the rich complexity of the real dynamic life situation.

Older views of the autonomic nervous system focused upon its peripheral structures and the belief that it functioned without awareness. It is now clear that the autonomic nervous system connections in the brain are rich and complex and extend into the highest cortical areas. We have also seen that the neurotransmitter substances that are involved in autonomic function are also found in the brain. The brain has important and essential roles in the function of the autonomic nervous system and, because autonomic structures have a certain level of intrinsic capability, we must not think that they are self-sufficient organs. As pointed out by Wolf (1975), the vital ability that is lost in a beating denervated heart is the ability to anticipate. We have hinted at the important relationships between movement and the autonomic nervous system but Malmo (1975) argues that the sole product of all the functions of the brain is concerned with the motor system. This is a defendable view but it falls a little short of being a balanced statement because, at the end of any argument of this nature, what appears to be the most important function will depend upon the primary interest of the person stating the case. We earlier referred to Carruthers and Taggert (1975) who have suggested that one of the most important roles of noradrenaline in the brain is to stimulate the 'pleasure' centres. They argue that physical exercise is the natural way of increasing noradrenaline levels in the brain but industrial and urban life have resulted in a diminished opportunity for increasing numbers of people to engage in physical activity. Carruthers and Taggert suggest that an important aspect of the man-and-his-environment interaction concerns the replacement of these drugs with nature exercise and relaxation techniques.

The techniques of biofeedback have demonstrated that some autonomic learning is a possibility. The hopes of the overoptimistic may not have been fulfilled but there is no reason why they should have been. Biofeedback methods are one of a number of useful techniques that can be used to help control autonomic malfunction.

Many autonomic functions evolved early in terms of the brain and in this sense they are archaic but, with each new phase in the evolution of the central nervous system, many autonomic functions were integrated into the evolving brain structures and altered in subtle ways. We have come to understand that the autonomic nervous system does not operate in terms of generalized discharges but like the brain in terms of discrete patterning. Many of the patterns involve the simultaneous discharge of systems that are usually thought of as operating in opposition. In many psychosomatic disorders we appear to be observing a lack of inhibition and we will not fully understand the significance of this overactivity until we learn to conceptualize the autonomic nervous system in terms of its central connections. The brain, the autonomic nervous system, and the endocrine system interact in complex cascades of function via the numerous chemical and anatomical pathways in the brain. It is the understanding of the dynamic qualities of these processes that constitutes the next step in our increasing knowledge.

REFERENCES

Ader, R. (1959). 'The effects of early experience on subsequent resistence to stress', *Psychol. Monogr.* (General and Applied), **73**, 1-32.

Adler, A. (1917). *Study of Organ Inferiority and its Physical Consequences*. Nervous and Mental Disease Publications, New York.

Alexander, F. (1950). *Psychosomatic Medicine: Its Principle and Application*. Norton, New York.

Alvarez, W.C. (1940). 'New light on the mechanisms by which nervousness causes discomfort', *J. Amer. Med. Assoc.* **115**, 1010-3.

Alvarez, W.C. (1951). *The Neuroses*. W.B. Saunders, London.

Appenzeller, O. (1970). *The Autonomic Nervous System: An Introduction to Basic and Clinical Concepts*. North-Holland, Amsterdam.

Bettelheim, B. (1959). 'Individual and mass behaviour in extreme situations', in E.E. Maccoby, T.M. Newcome and E.L. Hartley (Eds.). *Readings in Social Psychology (3rd edn.)*. Methuen, London.

Bovard, B.W. (1954). 'A theory to account for the effects of early experience on viability of the albino rat', *Science*, **120**, 187.

Burn, J.H. (1971). *The Autonomic Nervous System for Students of Physiology and of Pharmacology* (4th ed.). Blackwell, Oxford.

Calaresu, F.R. and Henry, J.L. (1971). 'Sex difference in the number of sympathetic neurons in the spinal cord of the cat', *Science*, **173**, 343-4.

Callaway, E. and Thompson, S.V. (1953). 'Sympathetic activity and perception: an approach to the relationships between autonomic activity and personality', *Psychosomatic Medicine*, **15**, 443-55.

Cannon, W.B. (1939). 'Homeostasis in senescence', *J. Mount Sinai Hospital,* **5**, 598-606.

Cannon, W.B. (1942). 'Aging of homeostatic mechanisms', in *Problems of Aging* (2nd ed.). Williams and Wilkins, Baltimore.

Carruthers, M. and Taggart, R. (1975). 'Man in noradrenaline secreting states'. Paper given at the Workshop on Catecholamines and Behaviour. (Convenor) M. Frankenhauser. Stockholm.

Denenberg, V.H. (1962). 'The effects of early experience', in E.S.E. Hafez (Ed.). *The Behaviour of Domestic Animals*. Williams and Wilkins, Baltimore.

Deutsch, F. (1922). 'Psychoanalyse und organkrankheiten', *Internationale Zeitschrift fur Psychoanalyse,* **8**, 290-306.

Dunbar, H.F. (1946). *Emotions and Bodily Changes* (3rd ed.). Columbia University Press, New York.

Eisdorfer, C., Nowlis, J. and Wilkie, F. (1970). 'Improvement of learning in the aged by modification of autonomic nervous system activity', *Science*, **170**, 1327-9.

Engel, B.T. (1960). 'Stimulus-response and individual-response specificity', *Arch. Gen. Psychiat.* **2**, 305-13.

Eppinger, H. and Hess, L. (1915). 'Die vagotonie', *Mental and Nervous Disease Monogr.* No. 20.

Eysenck, H.J. (1953). *The Structure of Human Personality*. Methuen, London.

Eysenck, H.J. (1968). *The Biological Basis of Personality*. Thomas, Springfield.

Folkow, B. and von Euler, U.S. (1954). 'Selective activity of noradrenaline and adrenaline producing cells in the cat's suprarenal gland by hypothalamic stimulation'. *Circulation Res.* **2**, 191-5.

Fortier, C., de Groot, J. and Hartfield, J.E. (1959). 'Plasma free corticosteroid response to faradic stimulation in the rat', *Acta Endocrinologica*l, **30**, 219-21.

Gitlow, S.E., Bertani, L.M., Wilk, E., Lan Li, B. and Dziedzic, S. (1970). 'Excretion of catecholamine metabolites by children with familial dysautonomia', *Pediatrics,* **46**, 513-22.

Graham, D.T. (1972). 'Psychosomatic medicine', in N.S. Greenfield and R.A. Steinbach (Eds.) *Handbook of Psychophysiology*. Holt, Rinehart and Winston, New York.

Gray, J. (1971). *The Psychology of Fear and Stress*. Weidenfeld and Nicolson, London.

Hanna, J.M. (1970). 'The effects of coca chewing on exercise in the Quechua of Peru', *Human Biology*, **42**, 1-11.

Heine, B. (Ed) (1972). 'Symposium: Life events and psychosomatic disorder', *J. Psychosomatic Res.* **16**, No. 4.

Henry, J.P. and Ely, D.L. (1976). 'Biological correlates of psychosomatic illness', in R.G. Grenell and S. Galay (Eds.) *Biological Foundations of Psychiatry*. Raven Press, New York.

Hess, E.H. (1972). 'Pupillometrics: a method of studying mental, emotional and sensory processes', in N.S. Greenfield and R.A. Steinbach (Eds.) *Handbook of Psychophysiology*. Holt, Rinehart and Winston. New York.

Himms-Hagen, J. (1967). 'Sympathetic nervous systems regulation of metabolism', *Pharmacol. Rev.* **19**, 367-461.

Hutchings, D.E. (1963). 'Early experience and its effects on later behavioural processes in rats III: effects of infantile handling and body temperature reduction on later emotionality', *Trans. New York Academy Sciences*, **25**, 890-901.

Jost, H. and Sontag, L.W. (1944). 'The genetic factor in autonomic nervous system function', *Psychosomatic Medicine*, **6**, 308-10.

Kempf, E.J. (1920). *Psychopathology*. Mosby, St. Louis.

King, F.A. (1958). 'Parameters relevant to determining the effects of early experience upon the adult behaviour of animals', *Psychol. Bull.* **55**, 46-58.

Kuntz, A. (1938). 'Histological variations in autonomic ganglia and ganglion cells associated with age and disease', *Amer. J. Pathology*, **14**, 738-95.

Kuntz, A. (1951). *Visceral Innervation and its Relation to Personality*. Thomas, Springfield.

Lachman, S.J. (1951). *Psychosomatic Disorders: A Behaviouristic Interpretation*. Wiley, New York.

Lader, M. (1975). *The Psycholphysiology of Mental Illness*. Routledge and Kegan Paul, London.

Lazarus, R.S. (1966). *Psychological Stress and the Coping Process*. McGraw-Hill, New York.

Levene, H.I., Engel, B.T. and Schulkin, F.R. (1967). 'Patterns of autonomic responsivity in identical schizophrenic twins', *Psychophysiol.* **3**, 363-70.

Levi, L. (1971). *Society, Stress and Disease*. Oxford University Press, London.

Levine, S. (1962). 'The effects of infantile experience on adult behaviour', in A.J. Bacharach (Ed.). *Experimental Foundations of Clinical Psychology*. Basic Books, New York.

Levine, S. (1966). 'Sex differences in the brain', *Scientific American*, **214**, 84-90.

Lipton, E.L., Steinschneider, A. and Richmond, J.B. (1965). 'The autonomic nervous system in early life. Parts 1 and 2', *New England J. Med*, **273**, 147-53 and 201-8.

McKusick, V.A., Norum, R.A., Farkas, H.J., Brunt, P.W. and Mahloudji, M. (1967). 'The Riley-Day syndrome: observations on genetics and survivorship', *Israel J. Med. Sci.* **3**, 372-9.

Malmo, R.B. (1975), *On Emotions, Needs, and Our Archaic Brain*, Holt, Rinehart and Winston, New York.

Mandler, G., Mandler, J.M. and Uviller, E.T. (1958). 'Autonomic feedback: perception of autonomic activity', *J. Abnormal Social Psychol.* **56**, 367-73.

Mason, J.W. (1971). 'A re-evaluation of the concept of "nonspecificity" in stress theory', *J. Psychiat. Res.* **8**, 323-33.

Mednick, S.A. (1970). 'Breakdown in individuals at high-risk for schizophrenia. Possible predispositional prenatal factors', *Mental Hyg.* **54**, 50-63.

Mednick, S.A. and Schulsinger, F. (1968a). 'Some premorbid characteristics related to breakdown in children with schizoprenia', in D. Rosenthal and S.S. Kety (Eds.) *The Transmission of Schizophrenia.* Pergamon, Oxford.

Mednick, S. A. and Schulsinger, F. (1968b). 'Some premorbid characteristics related to breakdown in children with schizophrenic mothers', *J. Psychiat. Res.* **6**, 267-91.

Nelson, R. and Gellhorn, E. (1958). 'The influence of age and functional neuropsychiatric disorders on sympathetic and parasympathetic function', *J. Psychosomatic Res.* **3**, 12-26.

Page, J. and Brown, G.M. (1961). 'Effect of heating and cooling the legs, hand, and forearm on blood flow in the Eskimo', *Applied Physiol.* **5**, 753-8.

Polikanina, R.I. (1961). 'Relation between autonomic and somatic components in development of conditioned defence reflex in premature infants', *Pavlovian J. Higher Nervous System Activity,* **11**, 51-8.

Rachman, S. (1960). 'Galvanic skin response in identical twins', *Psychol. Rep.* **6**, 298.

Rahe, R.H. (1975). 'Life changes and near future illness reports', in L. Levi (Ed.) *Emotions: Their Parameters and Measurements.* Raven Press, New York.

Ramey, E.R. and Goldstein, M.S. (1975). 'The adrenal cortex and the sympathetic nervous system', *Physiol. Rev.* **37**, 155-95.

Richmond, J.B., Lipton, E.L. and Steinschneider, A. (1962). 'Observations on differences in autonomic nervous system function between and within individuals during early infancy', *J. Amer. Acad. Child Psychiat.* **1**, 83-91.

Richmond, J.B. and Lustman, S.L. (1955). 'Autonomic function in the neonate: 1 implications for psychosomatic theory', *Psychosomatic Medicine,* **17**, 269-75.

Riley, C.M. (1952). 'Familial autonomic dysfunction', *J. Amer. Med. Assoc.* **149**, 1532-5.

Riley, C.M. (1967). 'Comments on familial dysautonomia', *Israel J. Med. Sci.* **3**, 355-7.

Roessler, R. (1973). 'Personality, psychophysiology and performance', *Psychophysiol.* **10**, 315-25.

Rubin, R.T., Gunerson, E.K.E. and Arthur, R.J. (1972). 'Life stress and illness patterns in the U.S. Navy VI: environmental, demographic, and prior life change variables in relation to illness onset in naval aviators during a combat cruise', *Psychosomatic Medicine,* **34**, 533-47.

Sarason, I.G., du Monchaux, C. and Hunt, T. (1975). 'Methodological issues in the assessment of life stress'. in L. Levi (Ed.) *Emotions: Their Parameters and Measurement.* Raven Press, New York.

Schaefer, T. (1963). 'Early experience and its effects on later behaviour processes in rats II: A critical factor in the early handling phenomenon', *Trans. New York Academy Sciences.* **25**, 871-87.

Schwartz, G.E. (1975). 'Biofeedback, self-regulation, and the patterning of physiological processes', *American Scientist,* **63**, 314-24.

Seligman, M.E.P. (1975). *Helplessness.* Freeman, California.

Selye, H. (1956). *Stress and Disease.* McGraw-Hill, New York.

Servais, J. and Hubin, P. (1965) 'Type neurovegetatif personalite et comportement psychopharmacologique', *Acta Neurology Belgique,* **65**, 152-68.

Shock, N.W. (1952). 'Ageing of homeostatic mechanisms', in *Cowdry's Problems of Ageing* (3rd ed.). Williams and Wilkins, Baltimore.

Smith, A.A. and Hui, F.W. (1973). 'Unmyelinated nerves in familial dysautonomia', *Neurology,* **23**, 8-11.

Stangl, E. (1952). 'Das fohnproblem in seiner beziehung zum vegetative nervensystem', *Wiener Klinische Wochenschrift,* **64**, 467-9.

Steinschneider, A. and Lipton, E.L. (1965). 'Individual differences in autonomic responsivity: problems of measurement', *Psychosomatic Medicine,* **27**, 446-56.

Steinschneider, A., Lipton, E.L. and Richmond, J.B. (1964) 'Autonomic function in the neonate: VI discrimination, consistency, and slope as a measure of an individual's cardiac responsivity', *J. Genet. Psychol.* **105**, 295-310.

Sternbach, R.A. (1966). *Principles of Psychophysiology.* Academic Press, London.

Stewart, K.D. and Dean, W.H. (1965). 'Perceptual, cognitive behaviour and autonomic nervous systems patterns', *Arch. Gen. Psychiat.* 12, 329-35.

Tursky, B. and Sternbach, R.A. (1967). 'Further physiological correlates of ethnic differences in response to shock', *Psychophysiol.* 4, 67-74.

Varni, J.G., Doerr, H.O. and Franklin, J.R. (1971). Bilateral differences in skin resistance and vasomotor activity. *Psychophysiol.* 8, 390-400.

Vollmer, H. (1927). 'Periodische pathothermie', *Zeitschrft fur Kinderheilk,* 43, 88-102.

Wallace, R.M. and Fehr, F.S. (1970). 'Heart-rate, skin resistance and reaction time of Mongoloid and normal children under baseline and distraction conditions', *Psychophysiol.* 6, 722-31.

Watts, G.O. (1975). *Dynamic Neuroscience: Its Application to Brain Disorders.* Harper and Row, New York.

Weiner, H. (1971). 'Current status and future prospect for research in psychosomatic medicine', *J. Psychiat. Res.* 8, 497-8.

Weininger, O. (1954). 'Physiological damage under emotional stress as a function of early experience', *Science,* 119, 285-6.

Weiss, J.M. (1971). I. 'Effects of coping behaviour in different warning signal conditions on stress pathology in rats'.

II. 'Effects of punishing the coping response (conflict) on stress pathology in rats'.

III. 'Effects of coping behaviour with and without a feedback on stress pathology in rats'.

J. Comp. Physiol. Psychol. 77, 1-30.

Wenger, M.A. (1948). 'Studies of autonomic balance in USAF personnel'. *Comp. Psychol. Monogr.* 19, No. 4.

Wenger, M.A. (1966). 'Studies of autonomic balance: a summary', *Psychophysiol.* 2, 173-86.

Wolf, S. (1963). 'A new view of disease', *J Amer. Med. Assoc.* 184, 129-30.

Wolf, S. (1975). 'Regulatory mechanism and tissue pathology', in L. Levi (Ed.) Emotions: Their Parameters and Measurement. Raven Press, New York.

Wolf, S. and Wolff, H.G. (1947). *Human Gastric Function.* Oxford University Press, Oxford.

Author Index

Numbers italicized indicate pages upon which the reference details can be found.

172

Schaefer, T., 151, *167*
Schapiro, S., 57, *73*
Scharrer, E., 57, *73*
Schildkraut, J. J., 103, *109*
Schlag, J., 68, *73*
Schneiderman, N., 38, *40*, 46, *74*
Schoenfeld, W. N., 113, *144*
Schwartz, G. E., 120, 122, *144*, 155, *167*
Seligman, M. E. P., 161, *167*
Selye, H., 39, *40*, 160, *167*
Servais, J., 155, *167*
Shapiro, D., 120, *144*
Sheehan, D., 7, *21*
Sherrington, C. S., 126, *144*
Sherry, L. V., 81, *109*
Shock, N. W., 152, *167*
Singer, C., 8, *21*
Sjarne, L., 97, *109*
Skinner, B. F., 113, 116, *144*
Smelik, P. G., 58, *74*
Smith, A. A., 149, *167*
Smith, K., 114, *144*
Smith, R. E., 37, *40*, 62, *74*
Smith, S. M., 117, *144*
Sokolov, E. N., 83, 84, *109*
Solomon, R. L., 127, *144*
Spalding, J. M. K., 117, *144*
Stangl, E., 154, *167*
Stanley, W. C., 69, *74*
Stanley-Jones, D., 132, *144*
Stein, L., 104, *109*
Steiner, G., 128, *144*
Steinschneider, A., 150, *167*
Steinwald, O. P., 100, *109*
Stellar, E., 53, *74*
Stern, J. A., 84, *110*
Stern, R. M., 118, *144*
Sternbach, R. A., 75, 81, 87, *110*, 131, *144*, 154, 159, *167*
Stewart, K. D., 157, *168*
Stohr, P., 32, *40*
Stroebel, C. F., 92, *110*
Strongman, K. T., 124, 135, *144*
Stunkard, A., 122, *144*
Surwillo, W. W., 83, *110*
Sutherland, A., 125, *144*

Tarpy, R. M., 112, *144*
Tart, C. T., 92, *110*
Taylor, N. B., 111, *144*
Teitelbaum, P., 55, *74*
Thornton, E. W., 116, *144*
Tobach, E., 125, *144*

Tomita, T., 32, *41*
Trendelenburg, U., 20, *21*
Trowill, J. A., 114, *144*
Tuke, D. H., 111, *145*
Tursky, B., 154, *168*

Udenfriend, S., 47, *74*
Ungerstedt, U., 49, *74*
Uno, T., 84, *110*
Usdin, E., 47, *74*

Valins, S., 136, 137, *145*
Vander, A. J., 89, *110*
Vanderwolf, C. H., 67, *74*
Van Toller, C., 128, 139, *145*
Varni, J. G., 155, *168*
Vassalle, M., 36, *40*
Venables, P. H., 75, *110*
Vogt, M., 47, *74*
Vollmer, H., 148, *168*

Wall, P. D., 68, *74*
Wallace, R. K., 92, *110*
Wallace, R. M., 148, *168*
Wang, G. H., 68, *74*
Warburton, D. M., 46, *74,* 92, *110*
Watson, J. B., 124, *145*
Watts, G. O., 63, 64, 65, 66, 67, *74*, 163, *168*
Webb, R. A., 89, *110*
Weil-Malherbe, H., 104, *110*
Weiner, H., 146, *168*
Weiner, J. S., 5, *21*
Weininger, O., 151, *168*
Weiskrantz, L., 129, *145*
Weiss, B., 62, *74*
Weiss, J. M., 161, *168*
Wenger, M. A., 78, 79, 80, *110*, 125, *145*, 150, 154, 156, *168*
Wenzel, B. M., 128, *145*
Wheatley, M. D., 135, *145*
Wilder, J., 81, *110*
Willshaw, D. J., 70, *74*
Wilson, E. O., 5, *21*
Winkler, H., 100, *110*
Wise, C. D., 104, *110*
Wolf, S., 148, 154, 164, *168*
Wooley, D. W., 49, *74*
Wurtman, R. J., 56, *74*
Wynne, L. C., 127, *145*

Zaimis, E., 128, *145*
Zimny, G. H., 84, *110*

Subject Index

Abel, 17

Acetylcholine, 6, 18, 32, 33, 36, 46, 93, 94, 104

Adrenal gland, 13, 14, 19, 33, 34–36, 38, 127, 129, 151
  adrenalectomy, 161
  cortex, 13, 39, 58, 61
  demedullation, 129
  medulla, 13, 19, 26, 60, 94, 129

Adrenaline (epinephrine), 17, 18, 19, 34, 36–37, 47, 60, 87, 98, 99, 100, 101, 102, 133, 136

Adrenergic (noradrenergic), 18, 62
  alpha adrenergic, 38
  beta adrenergic, 38

Adrenocorticotrophic hormone, 34, 39, 58, 59, 61

Ageing, 151–152, 157

Alcohol, 5, 87, 94

Amphetamine, 96

Amphioxus, 3, 4

Amylnitrite, 157

Anabolic, 20, 29

Animal spirits (vital spirits), 8, 10, 11, 17, 126

Anxiety, 99, 100, 104, 112, 113, 148

Aortic artery, 45, 46

Aristotle, 8

Arousal, 84–87, 137, 152, 156

Arthropoda, 2

Atropine, 18

Autonomic nervous system
  asymmetry, 155
  autonomic activity score, 158
  autonomic balance, 155
  autonomic fractionation, 87–90
  autonomic lability score, 83
  autonomic perception questionnaire, 122, 158
  autonomic scale, 78–81

climatic factors, 154
critical periods, 150–151
development, 150
idiosyncrasies, 3, 4, 154–158
personality variables, 155–159
social factors, 152–154

Autonomic tuning, 53, 78, 135

Avoidance learning, 120, 127

Basal metabolic rate, 39

Bernard, 12, 13

Bichat, 11, 12

Biofeedback, 7, 111–124, 139, 155, 159, 164
  animal studies, 114–116
  human studies, 116–124

Biogenic amines, 46–51

Blood pressures, 42, 44–46, 76, 77, 122, 155

Brain
  cerebellum, 68
  cortex, 68–70, 85, 124, 133, 134
  fourth ventricle, 13, 51
  hind brain, 43–46, 51
  hypothalamus, 13, 46, 51–55, 57, 58, 62, 65, 69, 70, 104, 119, 134, 135, 139, 151
  neomammalian, 66
  paleomammalian, 66
  pons, 43
  protoreptilian, 66
  reticular activating system, 43, 85, 134
  thalamus, 63–65, 70, 85, 133

Brown fat, 37, 61, 62

Brown–Sequard, 12

Carbon dioxide, 44

Carotid sinus, 28, 44, 45, 46, 89

Catabolic, 20, 28, 132

Catecholamine hypothesis of affective states, 93, 103–104